Communication for Accountants

Strategies for Success

MAURICE L. HIRSCH, JR.

Associate Dean for Academic Affairs
Professor of Accounting
Southern Illinois University Edwardsville

ROB ANDERSON

Professor of Communication
Saint Louis University

SUSAN L. GABRIEL

Corporate Writing Consultant

Boston   Burr Ridge, IL   Dubuque, IA   Madison, WI   New York
San Francisco   St. Louis   Bangkok   Bogotá   Caracas   Lisbon   London
Madrid   Mexico City   Milan   New Delhi   Seoul
Singapore   Sydney   Taipei   Toronto

*We dedicate this book to Janet Duthie Collins,
Isaiah Smithson, and our other colleagues in English,
Communication, and Business who have encouraged us
and worked with us over the years.*

Irwin/McGraw-Hill

A Division of The **McGraw-Hill** Companies

COMMUNICATION FOR ACCOUNTANTS

Copyright © 1999 by The McGraw-Hill Companies, Inc. All rights
reserved. Printed in the United States of America. Except as permitted
under the United States Copyright Act of 1976, no part of this
publication may be reproduced or distributed in any form or by any
means, or stored in a data base or retrieval system, without the prior
written permission of the publisher.

This book is printed on acid-free paper.

1 2 3 4 5 6 7 8 9 0 DOC/DOC 9 3 2 1 0 9 8

ISBN 0-07-038390-1

Vice president/Editor-in-Chief: *Michael W. Junior*
Publisher: *Jeffrey J. Shelstad*
Associate editor: *Rebecca M. Page*
Senior marketing manager: *Rhonda Seelinger*
Project manager: *Kari Geltemeyer*
Production supervisor: *Elizabeth LaManna*
Senior designer: *Crispin Prebys and Laurie Entringer*
Supplement coordinator: *Sue Lombardi*
Compositor: *GAC Shepard Poorman Communications*
Typeface: *10/12 Times Roman*
Printer: *R. R. Donnelley & Sons Company*

Library of Congress Cataloging-in-Publication Data
Hirsch, Maurice L.
    Communication for accountants : Strategies for success /
Maurice L. Hirsch, Jr., Rob Anderson, Susan Gabriel.
        p.   cm.
    Rev. ed. of: Accounting & communication.
    Includes index.
    ISBN 0-07-038390-1
    1. Communication in accounting. I. Gabriel, Susan L. (Susan
Laine), 1950. II. Anderson, Rob, 1945– . III. Hirsch, Maurice L.
Accounting & communication. IV. Title.
HF5625.5.H57   1999
302.2'024657—dc21   98-39390

http://www.mhhe.com

About the Authors

Maurice L. Hirsch, Jr., Professor of Accounting and Associate Dean for Academic Affairs at the School of Business, Southern Illinois University Edwardsville, has co-authored *Cost Accounting: Accumulation, Analysis, and Use,* Third Edition (South-Western, 1992), and authored *Advanced Management Accounting,* Second Edition (South-Western, 1994). He is also the author or co-author of several published cases and teaching notes. He is Case Section Editor of the *Journal of Accounting Education.* Along with Susan Gabriel, he has published several articles dealing with accounting and communication.

Rob Anderson, Professor of Communication, Saint Louis University, has worked with the other co-authors in developing a workshop sequence in communication for accounting students. Anderson has authored, co-authored, or co-edited seven previous books, including *Before the Story: Interviewing and Communication Skills for Journalists* (St. Martin's Press, 1989), *Questions of Communication: A Practical Introduction to Theory,* Second Edition (St. Martin's Press, 1998), and *Interviewing: Speaking, Listening, and Learning for Professional Life* (Mayfield, in press).

Susan L. Gabriel, Corporate Writing Consultant, was part of an ongoing program to involve readers from the English Department at Southern Illinois University Edwardsville in assessing the work of accounting students. She was the guiding force and prime facilitator for the workshop, "Writing Across the Business Curriculum," required for all School of Business faculty at Southern Illinois University Edwardsville. She has joined Hirsch in regional and national presentations on accounting and communication as well as collaborating on *Journal of Accounting Education* articles in this area. Gabriel has provided workshops and seminars for area businesses to improve corporate communication.

Preface

Accounting is an artifact—a product of human intervention designed by humans for human purposes.[1]

Purpose

When we began collaboration on this book, we had one goal in mind: to help you become a better communicator. Because accountants are required to relate information to others, oral and written communication are at the core of accounting. Thus, developing effective communication skills is essential for every accounting professional. Successful communication is achieved through careful thinking and planning; this book offers strategies for thinking through and developing a communication plan for a variety of communication contexts in the accounting field.

Accounting is primarily concerned with the accumulation, analysis, communication, and use of information. By its very nature, accounting (and the role accountants play) requires that its practitioners have excellent writing, speaking, reading, listening, and questioning skills. In the past, however, there has been an artificial division in how communication skills have been taught. Writing has been separated from speaking, and technical matters have been so separated from communication skills that eliciting information and listening skills have been generally ignored. In addition, the field has given insufficient attention to the clear link between communication and thinking skills. This book creates no artificial divisions and seeks to help you solve these problems.

As authors, we value all forms of communication that might be encountered by those in accounting-related positions as well as by managers in general. These forms include information presentation and reception both in oral and written formats, in one-to-one, one-to-many, and group discussion contexts. In addition, we believe thinking and communication are inseparable. Thus, our approach in this book is to address the convergence of the writing/speaking and thinking processes.

All intentional communication involves a series of choices, and making choices involves critical thinking. This realization focuses professionals on the creative control each person exercises over messages, even as a reader or listener. However, the writer or speaker has the extra responsibility to create communication plans that depend on a complex range of variables, including purpose, audience, role, and context. Individuals can develop thinking strategies to aid in this creation process and to make the entire process of communication less threatening and intimidating.

Intended Audience

We've written this book primarily for students majoring in accounting, for whom it can serve as reference and guide from the sophomore year through graduate school. Two other audiences will benefit: other undergraduate business majors, and practicing accountants (public, private, governmental) and managers. The book is a helpful supplement to all accounting courses and a useful reference guide for the practitioner's professional library.

What's Included and What's Not

The primary audience is students at the sophomore level and above who have completed their general course requirements in composition and oral communication. In addition, we assume students will use this book in conjunction with such professional resources as a dictionary, a thesaurus, and a writing/style manual; therefore, in matters of grammar and usage we concentrate on the larger communication issues that affect the clarity of accountants' messages. Beyond urging basic clarity, however, this book suggests new ways for accountants to make their communication more skillful and strategic. These larger communication issues include these essential topics:

- understanding the relationship of thinking and communication
- determining purpose
- analyzing an audience
- organizing a presentation
- preparing a draft
- editing a text
- working with a word processor
- dealing with stage fright
- overcoming writer's block
- speaking and writing with confidence
- interviewing
- listening
- communicating in formal and informal contexts
- coordinating group projects
- developing visual aids
- understanding nonverbal communication

Because readers will have varying skill levels, backgrounds, and experience, we have developed a wide range of examples and exercises. Obviously, not all examples or exercises will be equally accessible to every reader, but professionals and students alike should find explanations to be basic and practical.

General Structure

Our goal is to integrate material on writing and speaking skills rather than to have one chapter or section devoted to oral communication; face-to-face talk has been seriously undervalued in professional skills training. As appropriate, each chapter discusses common issues followed by how these same concerns can be specially addressed in speaking or writing situations.

We ask readers to consider examples of problems from both the business and academic worlds. These examples, labeled **Practice,** apply the chapter concepts to practical settings encountered by most accountants. Instructors can assign specific Practice exercises to be completed by their students; however, individual readers can benefit by considering and trying out the various Practice tasks, whether assigned or not. Suggested solutions and responses for each Practice site are included at the end of each chapter.

Because this is a supplemental book to be used in principles, financial, management, auditing, systems, and tax accounting courses, we encourage instructors to choose texts and cases that incorporate specific written and oral assignments as well as to redesign existing text problems and cases. The *Instructor's Resource Guide* expands on this point.

This book engages the reader directly with practical strategies and suggestions for how to apply concepts both to course work and on-the-job tasks. Since this is a guide, you can browse through it to find needed information. While there are some chapters that you will study from beginning to end, others will allow you to select only those parts that fit your needs.

Instructor's Resource Guide

The instructor's manual is a comprehensive guide to incorporating thinking and communication skills within accounting courses. Just as the book is an ongoing reference to its readers, the instructor's guide provides a resource to those who teach accounting. Instructors will learn how to choose different types of assignments, how to design assignments, how to give feedback, what to consider when grading, how to develop and utilize partnerships with English and communication faculty, what resources (technical, financial, human) are available, how the scope of overall course content is affected by integrating communication skills in a course, and how examinations and grade weights are affected.[2]

Acknowledgments

We wish to thank the following reviewers for their helpful evaluation, criticism, and suggestions during the development of this book: Michael Costigan, Southern Illinois University Edwardsville; Lynn Griffen, North Carolina Agricultural Tech; George Heyman, Oakton Community College; Debra Kerby, Truman State University; Linda Lovata, Southern Illinois University Edwardsville; Patti Mills, Indiana State University; James S. O'Rourke, University Notre Dame; Andrew Rosman, University of Connecticut; Barbara W. Scofield, University of Texas of the Permian Basin; Fred Sellers, Southwestern University; James A. Smith, Saint Cloud State University; David E. Stout, Villanova University; Ronald Tidd, Michigan Technological University; Loren Wenzel, Mankato State University; Donald E. Wygal, Rider University.

We also want to thank some wonderful people at Irwin/McGraw-Hill. First is Becky Page, who has been excited about this project from the first time she saw it. Her leadership has been very helpful throughout this process. In addition, our project manager, Kari Geltemeyer, helped us stay organized and on track. We had two very talented designers on the project, Crispin Prebys and Laurie Entringer. Finally, we want to recognize Jeff Shelstad, our Publisher, for his enthusiasm and encouragement.

Bud Hirsch
Rob Anderson
Susan Gabriel

Endnotes

1. L. Parker, K. Ferris, & D. Otley, *Accounting for the Human Factor* (Sydney: Prentice-Hall of Australia Pty Ltd., 1989), p. 1.

2. Much of the basic philosophy of the instructor's manual comes from S. Gabriel and M. Hirsch, "Critical Thinking and Communication Skills: Integration and Implementation Issues," *Journal of Accounting Education* 10.2 (Fall 1992): pp. 243–270; and "Feedback Strategies: Critique and Evaluation of Oral and Written Assignments," *Journal of Accounting Education,* 13.3 (Summer 1995): pp. 259–279.

Contents

Accounting and Communication

66 **Being the best qualified is not enough. Applicants must be able to communicate their qualifications ... That's why it's crucial for applicants to ... communicate effectively.** **99**

PRICEWATERHOUSECOOPERS
HOME PAGE, TOOLS
FOR SUCCESS

Chapter 1
Communicating in Accounting

"TRANSLATE THIS EASY-to-READ STATEMENT INTO JARGON, WILSON, So A COMPANY WILL HAVE to HIRE AN ACCOUNTANT to TRANSLATE IT."

Imagine you have been asked to explain the field of accounting to a friend who has a liberal arts degree, a group of high school students on career day at the school, and a client who is an entrepreneur opening a restaurant. You have a sense of what you want to say about accounting, but you are faced with a series of decisions in each situation regarding your role, the audience you are addressing, the task at hand, and the ways you will communicate your message. Some of the interactions will be informal, while some should be formal. You must make choices about what should be stated orally and what should be put in writing. Within both oral and written presentations, there are additional decisions about format (e.g., what in words and what in charts? what in lists and what in full sentences?). In this book we suggest strategies to help you make such communication decisions.

When explaining accounting to those outside the profession, you might not think to emphasize that accounting is communication. However, *accounting is the accumulation, analysis, and coordinated use of information; and the communication of relevant information to decision makers.* While much of accounting deals with financial information, limiting its definition to just this part of accounting is too narrow. As an accountant you may communicate with current or potential stockholders, financial analysts, banks or other financial institutions, public accountants, and internal managers, among others. Your role can change, as the example at the beginning of the chapter illustrates. There are many specific tasks to consider. You might be writing a letter of engagement, giving a presentation about some tax research you have just completed, or both discussing and presenting a written report about the internal controls at a client company. You could be making an oral presentation to the next level up in your organization about a proposed business plan a team of you and your co-workers have been working on for three months. Your audiences can include very knowledgeable people who understand technical jargon as well as those who are less informed and who need information presented in plain English. In addition, your communication involves both sending and receiving messages, eliciting and presenting information, and choosing what oral or written formats are appropriate.

This book helps you plan and analyze your communication whether you are working alone or in groups. By considering a variety of issues, you will be better able to develop your own style of communication. In addition, accountants should value aspects of professional communication that are often neglected: reading, listening, and nonverbal sensitivity. Many books and articles advise or give tips and techniques to try. This book goes beyond simple techniques to provide a more complete communication framework with which to achieve greater success on the job. In this first chapter you will consider the relationship between accounting and communication, types of communication, the link between thinking and communication, and what skills are necessary to be able to communicate effectively.

Communication: The Core of Accounting

Accountants are both preparers and users of information. Information is news; it is useful to a decision maker and reduces uncertainty about the environment. As a preparer of information, you might be writing notes to accompany financial statements, interpreting auditing standards for a client, communicating with the Internal Revenue Service about a tax question, developing a new company's business plan that will go to lending institutions, or presenting an analysis of a possible capital asset acquisition to operating managers. Thus, communicating can involve all aspects of accounting (financial, managerial, tax, auditing, and systems) addressed to people who are either internal or external to the organization. Accountants can be part of the management team or can be consultants to management. In all cases, though, it is your responsibility to communicate effectively and efficiently with your audience, given the task at hand.

ROLE OF ACCOUNTANTS

In the past, accountants might have seen their role merely as advisers, in staff rather than line positions. In such positions, all that is required is to present information and leave it to others to interpret and make decisions. For example, one of our fathers was an accountant during the 1950s at Ford Motor Company. Accounting personnel at his plant were expected to prepare reports for others to use and explain. They were not at management meetings; they were isolated from operations. In their careers, they might have never written a memorandum other than a technical note for the files. They were seen as number crunchers. This role has changed quite a bit over the years. Operating managers are telling accountants that they want not only "the figures" but also a complete analysis. In addition, managers might expect accounting personnel to participate in operating decisions. The role of CPA firms has changed. Besides performing audits and doing tax research, more and more firms are providing consulting services. In most cases, this means more than just providing background information; clients want conclusions and recommendations. Thus, accountants both prepare and use information as an integral part of the management decision-making team.

YOU CANNOT AVOID COMMUNICATING

Some people imagine that it's just not that important to develop good critical thinking and oral–written communication skills. You might think that entry-level positions just require good technical skills in accounting. Looking down the road is too far off. And don't you just need to be a good accountant to progress? It's important that you realize the need to develop strong critical thinking and communication skills and that if you want to progress in any organization, you must demonstrate these skills right away and strengthen them even more. Recruiters will often bypass someone who is technically able but who does not present information well. Once in a company, oral and written reports are a way your supervisors can assess who has potential and who does not. Even if you are in a job where you mainly interact with a computer doing research, you will have to communicate with others. Your boss will want you to establish goals and to assess how well you are achieving them. Others will want e-mail or written reports outlining the results of your work. You cannot avoid communicating with others. You will have to sell ideas, motivate others, and be assessed by others. Communicating is a condition of employment and a key to your success as an accountant.

WAYS WE CAN COMMUNICATE

In some sense, modern life is plagued by too many communication choices. You can generate hard copies of letters, memoranda, and spreadsheets; send electronic mail to be viewed on a screen; record voice mail messages; produce CD-ROM, DVD, or videocassettes; have direct phone conversations; and meet one-on-one or in groups either formally or informally, in unmediated or mediated situations. To gather information, you have the same sources plus access to the Internet, international data banks, and highly specialized library documents. When presenting and eliciting information, you make many choices. This book develops a framework for communication so your choices are both more informed and explicit.

WHAT THE BIG SIX SAY ABOUT COMMUNICATION

A review of the home pages of the Big Six firms shows that communication skills are a key ingredient of landing a job and providing value-added client services. Arthur Andersen, in discussing career opportunities, states the following about the financial audit assurance process:

> *Communication is an ongoing process. In client meetings, both formal and informal, we confirm our understanding of audit issues, risk assessments, process flows, and other business issues. We communicate our findings, frequently using visual presentations to clarify our points. [http://www.arthurandersen.com/homepage.asp]*

PricewaterhouseCooper states the following as part of their Tools for Success:

> *Today's employers are seeking the best-qualified individuals for their limited number of openings. Being the best qualified is not enough. Applicants must be able to communicate their qualifications to the interviewer. That's why it's crucial for applicants to be prepared—and to communnicate effectively.* [http://www.pwcglobal.com/us/eng/car-inexp/workwithus/tools.html]

Ernst & Young LLP answers the question "Do you have what it takes?" by stating:

> *Communication. Are you effective, clear and able to get your point across? [http://www.ey.com/careers/newhire.asp]*

Types of Communication

You are the chief financial officer of a new company that is going to buy an ongoing manufacturing business from its current owners. Over the last few months you have been working with several banks and other financial institutions to finance the start-up of the company. This involves buying the firm from its existing owners and developing funds for capital expansion and for working capital. You are at the point where two financial institutions have presented offers to your company. Neither is a perfect fit, and there are trade-offs between them as well as many points to be negotiated before a final decision is made. You are now deciding what information the finance committee of the new company's board of directors should have and how to communicate it. This committee consists of the chairperson of the board (a major investor with a service industry background), the president (currently in charge of the company that your new firm will be buying), and two members of the union that represents the workers in the plant.

Consider the choices that you have in this situation. You can present financial information in summary or in detail, in tabular or graphic form with or without accompanying notes. You can hand out copies of information and/or use overhead transparencies or computer-generated slides. Some information can be written, and some can be spoken. You must also decide how technical you want to be. You have to focus on a primary objective of the presentation—stressing the need for a decision today, introducing the issues to the audience, or persuading the board to accept your conclusions.

This problem illustrates the multiplicity of communication choices in accounting. Your options include communicating in writing or orally, formally or informally, and verbally or nonverbally. You might be communicating with an individual or with a group. You might be listening/receiving, actively eliciting information, or presenting it to others. You must make informed and explicit choices about how you are communicating. Your goal is to develop communication that works best for all the parties concerned and helps them focus critically on relevant information.

We discuss written and oral communication in this book. When we describe *written communication,* we include any form of communication that involves creating a text to be read by the audience. It might be a handwritten letter, a computer-generated memorandum, or electronic mail. *Oral communication* includes not only face-to-face situations but also many electronically mediated messages, even those in which the communicators are removed from each other in time and space. Oral communication is usually direct interpersonal talk, but also could include teleconferencing, training videos, messages on voice mail, and telephone calls. Even some written forms, such as e-mail, may exhibit certain characteristics of spoken language. Many accounting tasks involve a combination of written and oral messages. Now, let's move to another important link: thinking and communication.

Critical Thinking and Communication: An Inseparable Combination

Effective organizational communication involves choosing among available options. Making these choices requires critical thinking. *Critical thinking* is the process of systematically investigating and evaluating alternatives for solving problems. It involves both *analysis* (separating a problem into its component parts) and *synthesis* (recognizing how the parts of a problem relate as a whole).

In written communication, for example, you start out with a blank piece of paper or computer screen. You create text through a series of choices—general organization, format, major sections, paragraphs, sentences, words. You are in control of the decision-making process, but the choices you make depend on a string of variables including purpose, intended audience, and your role. In writing this book, we had to think through these dimensions and keep rethinking them as we wrote each chapter, section, paragraph, sentence, and word. By becoming aware of the full range of decisions to be made, you are less likely to be trapped in oversimplified assumptions. Understanding how complex communication can be actually makes it *less* threatening and intimidating.

If asked, many people will say they are afraid of getting up and speaking in front of others; however, most people will also say they have much more trouble with written communication than oral. If you are one of those who dislike writing or have problems with producing good written communication, it may help you to consider that as a writer, you have virtually complete control over the creation

Levels of Critical Thinking

Consider the following tasks. How are they the same and how are they different in the level of critical thinking required? All involve a financial accounting standard (FAS) created by the Financial Accounting Standards Board (FASB).

Task 1: Write a synthesis of FAS 131 for an audit client.

Task 2: Write a synthesis of FAS 131 for an audit client and include an analysis of how her company is affected by this standard.

Task 3: Write a synthesis and analysis of the arguments that accounting firms and industry made before FAS 131 was adopted. Include your recommendation on the position your firm should take in this argument and how that position should be expressed.

(writing) process. No reader will know how many drafts you wrote; no one will be aware of the number of times you rewrote your first sentence. Knowing you have control can be an overwhelming responsibility, but it can also be empowering.

People who have strong critical thinking skills also must realize that substance (the result of their analysis) is inseparable from the form of communication. Critical thinking is an integral part of communication; good critical thinking is useless if the results of that analysis cannot be communicated to others. Similarly, something that is grammatically correct and contains all the right buzzwords but lacks any depth of analysis is equally useless to managers who want information for decision making. It is false to assume that the mechanics of writing or speaking can be separated from a concern for content. Our view is that *critical thinking and communication are inseparable.*

Strategies for Clear Thinking

Effective thinking never comes automatically. Applying the following strategies will enable you to develop a well-thought-out communication plan. *Come back to this section and refer to it as you start thinking about a new task.*

1. **Focus on the problem or task.** Before you can analyze a situation, you need to free yourself from outside distractions and concentrate on the task at hand. Put yourself in an environment that encourages uninterrupted thinking. This might mean completing other tasks (e.g., returning telephone calls, paying bills) before starting.

2. **Determine your goals or objectives.** Part of the skill of focusing is defining your problem clearly and accurately. What, specifically, needs to be addressed? Decide precisely what you are hoping to accomplish. Too often, if you have trouble organizing your presentation, or maintaining your focus, you do not have a clear idea of the major goal you want to convey. If

---

STRATEGIES FOR CLEAR THINKING

- Focus on the problem or task.
- Determine your goals or objectives.
- Consider the big picture.
- Follow each idea or solution to its logical conclusions.
- Consider each idea from an alternative angle or point of view.
- Avoid reactive thinking.
- Take a break before the final assessment of your ideas and solutions.

---

you are not clear about your goals, then you have no way of developing a meaningful strategy and no way of assessing the eventual outcome. For example, consider the possible goals in the finance committee scenario described earlier.

3. **Consider the big picture.** Too many people fail to think about all the factors that make up any given situation—for example, how does my decision here affect others? If you narrow your viewpoint too much, you may miss important aspects that are relevant to the final outcome. Effective problem solvers remember that both analysis and synthesis are crucial. They also consider how a specific piece of communication fits within a larger context and set of objectives. For example, an initial report might develop some background material while follow-up reports present conclusions and recommendations. Knowing the goals in a more global sense is important in developing the objectives of each piece of communication leading up to a planned conclusion.

4. **Follow each idea or solution to its logical conclusions.** For every idea you develop, take the next step and ask yourself: "If I do this, what is likely to happen?" Actions have consequences; thus, any decision you make must be considered in the context of probable results.

5. **Consider each idea from an alternative angle or point of view.** By putting yourself in the place of another person, you can often see a situation creatively. If you were the intended audience, what would you think and how would you feel about a particular set of recommendations? While an audience might not always agree with you, you want to present your arguments in a way that will either (*a*) appeal to listeners/readers or (*b*) not alienate them. Remember that *ideas* are not inherently appealing or unappealing. *People* are inherently evaluators of ideas from their own perspectives. You need to understand those perspectives. This alternative viewpoint can help you to better assess the problem or situation and may even help you to arrive at a more acceptable solution.

6. **Avoid reactive thinking.** When you make a split-second judgment or decision without considering the ramifications of your actions, you are

PRACTICE

Clear Thinking

Your boss comes into your office and says, "I just returned from a meeting where they talked about how we could save money by doing our own tax returns rather than having an outsider do them. I want you to prepare a presentation for senior management to explore and explain this idea."

Task: Use the steps above to plan an approach to this request.

---

guilty of *reactive thinking,* thinking that comes strictly as a reaction to a problem. For example, you might feel pressure from your supervisor or client. Therefore, you take some action to relieve the pressure. Reactive thinking may lead to solutions that work in the short run, but often this type of thinking can cause long-term problems because you did not take time to assess carefully the likely aftermath of an action. Reactive thinking is a particular problem in spur-of-the-moment responses to questions or e-mails, but beware of its effects also when sitting down to write a letter or report or when preparing specific comments for an oral presentation.

7. **Take a break before the final assessment of your ideas and solutions.**
   You will be able to make clearer judgments about your ideas if you put the work aside for a while. This gives your brain a break and allows you to return to the problem with a fresher, more objective outlook. If possible, give yourself a day or two before you finally go back to assess your ideas. If your schedule does not allow you the luxury of a day, then take a break and allow yourself to think of other things before returning to your task. In order to take a break, however, you must plan ahead and avoid doing a task at the last minute.

Clear thinking is the first step to clear communication.

Necessary Communication Skills

Communication is required as a normal part of all jobs, a fact commonly reinforced by recruiting manuals and upper-level management. For many years practitioners have sent a strong message to those who are about to enter the profession: *Technical accounting skills are not enough; accountants must be able to communicate effectively through both writing and speaking.*

Thirty percent of the accounting firms surveyed by *The Wall Street Journal*[1] reported dissatisfaction with the communication skills of entry-level accountants. Another study[2] cited poor writing skills as a major reason for job terminations among entry-level accountants. Stout and associates[3] report that practicing CMAs place great importance on communication skills (especially speaking, writing, and listening). Novin and Pearson[4] surveyed partners in CPA firms to find out, among other things, the skills these partners thought were necessary for entry-level public accountants. Novin and colleagues[5] sought the same information regarding entry-level management accountants from controllers of United States corporations who are CMAs. Respondents to these two studies thought the following skills were either very important or extremely important for entry-level accountants:

	CPA Partners	**CMA Controllers**
Thinking skills	96%	95%
Problem-solving skills	95	97
Listening skills	92	91
Writing skills	89	92
Speaking skills	77	49

Novin and colleagues conclude that:

> Writing, listening, verbal, and overall management skills were identified as either important skills . . . or as current weaknesses. . . . The need for these skills indicates that more practice in public speaking, writing, and group discussion is warranted in all phases of the accounting curriculum.[6]

The need for better communication and critical thinking skills is also emphasized in the report of the American Accounting Association Committee on the Future, Structure, Content, and Scope of Accounting Education (the Bedford Committee),[7] by the (then) Big Eight accounting firms[8] and by the Accounting Education Change Commission (AECC).[9] The Big Eight statement on accounting education claims:

> Public accounting requires its practitioners to be able to transfer and receive information with ease.
>
> Practitioners must be able to present and defend their views through formal and informal, written and oral, presentation. They must be able to do so at a peer level with business executives.[10]

The AECC states:

> To become successful professionals, accounting graduates must possess communication skills. . . . Communication skills include both receiving and transmitting information and concepts, including effective reading, listening, writing, and speaking.[11]

Among other things, in order to become skilled in communication and successful in accounting, you have to work on thinking, writing, reading, speaking, and listening. Technical competence, however, is merely the first step to accounting success. Even if you can communicate effectively through writing, if you cannot be an active participant in a problem-solving discussion with your peers and/or supervisor, or if you are unable to address a large group, you have limited your own growth and potential. Others at your organization will notice, too, how you are limiting the organization.

It's scary to think about all these skills. Certainly most of us are more comfortable with particular styles of communicating, like informal conversation or creating computer spreadsheets. Most of us will never be equally comfortable with all the dimensions of communication. Some will enjoy public speaking, while others will hate it. Some will appreciate the challenge of writing lengthy analyses, while others will cringe at the thought. However, the objective of communication training is to put you more in control of your own destiny and direction. While you might not develop excellent skill levels in all the areas explored in this book, you will at least be more relaxed and competent if you follow these guidelines.

BASIC SKILLS ASSUMED

If you are using this book in conjunction with a university course or are a practicing professional, we assume that you have taken basic English and speech communication classes. We also assume that you will use this book along with an appropriate reference writing/style manual. While we do cover some basic items you might find in a style manual (e.g., see Chapter 6), we believe that it's important to have and refer to a complete stand-alone manual separately. See *Would You Like to Know More?* at the end of this chapter for both Internet and hard copy style references. Therefore, we will concentrate on the most important issues that affect practicing accountants in addition to some basic language conventions such as grammar and punctuation.

YOU MUST BE AN ACTIVE LEARNER

We want to engage you directly in the learning process. Each chapter includes practical, viable strategies and suggestions that reflect our recognition of what accountants actually *do*. You can apply the concepts both in course work and on the job. You can use this book as an organized program of study by reading through it from beginning to end, and can also use it for troubleshooting by investigating specific answers to questions and problems you encounter. It is important to see that we are not proposing mere theory here or something that you can gain only through "book learning." Becoming skilled in communication involves ongoing practice so that competence becomes second nature. Here are some keys to active learning.

- **Inquire and clarify.** Whether you are completing a communication task as part of a course or for a job, it's up to you to ask questions that help you to clarify the basic objective of a presentation; your role, audience, and task; and the form and format of your end product.

- **Set up an appropriate schedule for the project and adhere to it.** Planning in advance and following the plan are important to achieving your desired ends. This also allows time for clarification and redrafting.

- **Draft and redraft.** We discuss this in more detail later in the book, but whether you are preparing an oral or written presentation, you should create complete drafts to have others listen to or review.

- **Participate in peer reviews.** Active learning also involves listening to and reading others' drafts. You can pick up pointers that will help you (style, content, etc.), and you can also see the things that do not work. This helps you as well as the person who has asked for your input.

WHAT'S INCLUDED IN THIS BOOK

The rest of this book is divided into three major sections.

1. **Developing a communication plan.**
   —Chapter 2 describes how to analyze the context and purpose, establishing the idea that all communication involves coordinating roles, audiences, and tasks. Communication is not just a neat sequence of information transfer, but a delicate process of collaboration.
   —Chapter 3 emphasizes the importance of analyzing your audience as you prepare to communicate.
   —Chapter 4 helps you develop your actual content.

2. **Implementing a communication plan.**
   —Chapters 5 and 6 discuss how to create (draft) a persuasive text and polish it through editing and revising.

3. **Applying practical ideas to focused communication contexts.**
   —Chapter 7 suggests strategies for writing memoranda, letters, e-mail, proposals, reports, cases, and group writing projects.
   —Chapter 8 helps you respond to special oral communication challenges, including nervousness or stage fright, interviews, presentations, and group collaborations.

Except for the last two chapters that analyze special problems associated with written or oral communication formats, all other chapters include relevant elements of both formats and at times discuss listening and eliciting of information as well.

Summary

Accounting involves the sharing of relevant information and interpretations among decision makers. As managers or accounting-related professionals, you are faced with many choices on how to communicate effectively given your role, the audience you are addressing, and the task at hand. You might be more comfortable with some ways of communicating than with others. However, by taking a critical look at how to communicate better orally and in writing, formally and informally, to individuals and to groups, and by developing listening and information elicitation skills, you can increase your chances for success and take more control over your own career.

How to Approach Practice Tasks

LEVELS OF CRITICAL THINKING

The important words in this assignment are *synthesis, analysis,* and *recommendations*. An important context is the shift in audiences: Tasks 1 and 2 have your firm's audit client as the audience, while Task 3's audience consists of technical people in your accounting firm. FAS 131 is a good example to use since it involves some real issues for most companies on segment reporting.

- **Synthesis** involves a summary of the salient parts of the FAS itself and relating them to each other. The objective here is to give the intended audience a clear and concise understanding of what the standard is all about. Language should reflect terms the audience understands. This usually means demystifying for your audit client jargon terms that often appear in official pronouncements.

- **Analysis** requires that you go beyond merely describing the content; you need to view the material in terms of its relevance to a particular context or a particular set of criteria. Effective analysis goes beyond pointing to the content and creatively dissects a problem to see what makes it tick. This means asking **why?** and **how?** and **so what?** You will also have to make choices in form and format for your analysis: What do you want to put in words, in tables, in charts or graphs? In Task 2, your analysis is directed toward an understanding of how a client is affected by the standard. Perhaps there is little effect since the new rule applies to something that does not relate in any material way to your client's firm (e.g., they have only a single line of business). On the other hand, the standard could mean significant changes for the client, and these must be explained. Some of the issues might be charged with emotion, given the past reporting practices of the client firm and the way its managers think of their business as compared to the way the standard asks them to think about it.

- **Recommendations** (or **conclusions**) go beyond analysis. In your synthesis you are summarizing. In your analysis, you are discussing implications. You are also building both argument and evidence that lead to the final stage of making recommendations. Any conclusions you reach should flow naturally from what precedes them. While your audience might not always agree with your recommendations, readers (and listeners) should be able to understand how you arrived at them and how they are a valid extension of your synthesis and analysis. In the case of this standard, where different interpretations are possible and past practices of the client might be affected, building the basis for the recommendations is very important, as is couching your conclusions in a way that will be considered favorably by the client.

CLEAR THINKING

As with most of the hypothetical situations described in this book, the description does not give enough information for definitive answers. Therefore, instead of looking at a specific solution, use the details you do know about the task to probe or test your understanding of the seven suggested critical thinking strategies.

1. Defer the planning of your response if possible until you can give it your full attention.

2. In order to clarify your own goals, think carefully about the wording and presumed intent of the supervisor's request. Is he or she asking for a presentation that would *advocate* internal tax return preparation, or one that would simply *describe* the advantages and disadvantages of such a change? Once you know what you hope to accomplish, you can decide on the best format for your presentation.

3. Place this single task in the "big picture" of the organization by asking yourself whose positions would be helped, harmed, changed, or bolstered by the changed procedures. You'd probably need specific data about how much is paid currently to outside preparers, how accurate their work has been, and the existing organizational expertise and potential resources available to commit to this task. Also consider short- versus long-term trade-offs of the decision. This information will allow you to generate some alternative solutions to your task.

4. Anticipate possible outcomes of your alternative choices. For example, if you prepare a written executive summary comparison, what should be its length and what are its chances of being read carefully? If you ask for a brief meeting to explain the alternatives to senior management, how likely are they to allot the time you need to do valid comparisons?

5. As much as possible, identify or empathize with the people who will be affected by who prepares the tax return or who will implement the decision. Think about their concerns and what points they would think were important. Anticipating outcomes realistically and imagining reactions are excellent ways to begin to evaluate alternatives in problem solving.

6. Strategies six and seven are checks on the process that could save you from rash decisions. Consider your own personality and previous work habits to make sure you do not jump to an early conclusion that keeps you from considering other effective solutions. For instance, your first inner reaction to the boss's statement might be "That's stupid. Why spread ourselves even thinner in this department when Talbot, Morslick have been doing such a great job on taxes?" This dismisses the suggestion too soon, and even though it's an internal reaction, it will affect subtly the commitment you feel to the presentation. In the same way, don't let the interest of your boss influence your thinking. Your job is to investigate; it's senior

management's job to make final judgments. Don't confuse the two roles.

7. Finally, suppose you've decided to make a balanced presentation of the pros and cons of internal tax preparation. Once you have a sense of how you're going to proceed, put the project aside for a while. When you come back to it, review your ideas and intended approach. You might have more or better ideas. Some of your strategies might look less promising. Make any revisions in your plan you think are necessary.

Obviously, these strategies will not necessarily make your presentation easier or magically make your decisions for you. They may, however, clarify matters enough to make your decisions more helpful to your organization.

Appendix:
Would You Like to Know More?

If you believe that you need additional background in basic communication skills, refer to one or more of the following.

CRITICAL THINKING

These three books represent recent thought about the importance of critical thinking, both in education and in professional life.

Corbett, E. *The Elements of Reasoning.* New York: Macmillan Publishing Company, 1991.

Paul, R. *Critical Thinking: What Every Person Needs to Survive in a Rapidly Changing World.* Sonoma, CA: Center for Critical Thinking and Moral Critique, 1990.

Siegel, H. *Educating Reason: Rationality, Critical Thinking, & Education.* New York: Routledge Chapman & Hall, Inc., 1988.

Other books in this area include the following.

Ruggiero, V. *The Art of Thinking.* New York: Harper & Row, 1988. A discussion of fallacies of reasoning in everyday life; Ruggiero provides many vivid and interesting examples.

Hollihan, T. A., and K. T. Baaske. *Arguments and Arguing: The Products and Process of Human Decision Making.* New York: St. Martin's Press, 1994. These authors merge critical thinking and argumentation principles, encourage readers to have more productive public interactions with others.

WRITING AND READING

For basic skill guidance and inspiration in the field of written communication, you may want to refer to the following.

Adler, M. *How to Read a Book.* New York: Simon & Schuster, 1940. This is a classic description of what a reader can do to enhance communication with an author.

Elbow, P. *Writing Without Teachers.* London: Oxford University Press, 1973. A creative and entertaining book. Elbow justifies how each of us can become writers who are reader-centered.

Fuess, B., Jr. (Ed.). *How to Use the Power of the Printed Word.* Garden City, NY: Doubleday Anchor Press, 1985. This award-winning series of International Paper Company ads appeared in newspapers and national magazines early in the 1980s. It includes practical tips on punctuation, style, vocabulary, and other issues from such noted writers as Russell Baker, Erma Bombeck, John Irving, Jane Bryant Quinn, and Kurt Vonnegut.

Goldberg, N. *Writing Down the Bones: Freeing the Writer Within.* Boston: Shambhala, 1986. Do you get stuck or blocked as a writer? Goldberg's suggestions help you decide when, where, and in what mood to write, not just how to write.

O'Conner, P. T. *Woe Is I: The Grammarphobe's Guide to Better English in Plain English.* New York: G. P. Putnam's Sons, 1996. A clever and entertaining source for exploring grammar and punctuation problems all writers must face.

Schwartz, M., and the Task Force on Bias-Free Language of the Association of American University Presses. *Guidelines for Bias-Free Writing.* Bloomington, IN: Indiana University Press, 1995. This book provides excellent reminders for authors who do not want to offend readers unintentionally, and for all of us who simply want to expand our sensitivity to other social and cultural groups.

Strunk, W., Jr., and E. B. White. *The Elements of Style* (3d ed.). New York: Macmillan, 1979. This is the best-known guide to writing style. Though it adopts an authoritarian tone at times, many writers still swear by it.

Zinsser, W. *On Writing Well: An Informal Guide to Writing Nonfiction* (6th ed.). New York: Harper & Row, 1998. Zinsser is considered a writer's writer. He includes especially effective discussions of editing, along with an excellent chapter on working with a word processor.

SPEAKING AND LISTENING

The following sources provide excellent discussions of the basis of speaking and listening, especially in professional settings.

Bolton, R. *People Skills.* Englewood Cliffs, NJ: Prentice-Hall, 1979. This is perhaps the best popularized introduction to the range and importance of interpersonal communication skills.

Hoff, R. *"I Can See You Naked": A Fearless Guide to Making Great Presentations.* Kansas City: Andrews and McMeel, 1988. Although this book has a silly title, it is amazingly practical. It includes excellent strategies for alleviating communication apprehension and stage fright.

Purdy, M., and D. Borisoff. *Listening in Everyday Life: A Personal and Professional Approach* (2d ed.). Lanham, MD: University Press of America, 1997. A thorough and practical look at an often-neglected aspect of communication. This book will help you listen across differences of gender, culture, and value.

Stone, J., and J. Bachner. *Speaking Up: A Book for Every Woman Who Wants to Speak Effectively.* New York: McGraw-Hill, 1977. An exploration of the special challenges and opportunities that women speakers face in organizational life.

Endnotes

1. S. Feinstein, "Words Count," *The Wall Street Journal* (July 15, 1986), pp. 1, 5.

2. *Management Accounting Campus Report,* May 1989 (Montvale, NJ: Institute of Management Accountants).

3. D. Stout, D. Wygal, and R. Hoff, "Writing across the Disciplines: Applications to the Accounting Classroom," *The Bulletin of the Association for Business Communication* 53.4 (1990), pp. 10–16.

4. A. Novin and M. Pearson, "Non-Accounting Knowledge Qualifications for Entry-Level Public Accountants," *The Ohio CPA Journal* (Winter 1989), pp. 12–17.

5. A. Novin, M. Pearson, and S. Senge, "Improving the Curriculum for Aspiring Management Accountants: The Practitioner's Point of View," *Journal of Accounting Education* 8 (1990), pp. 207–224.

6. Novin et al., p. 220.

7. American Accounting Association [AAA] Committee on the Future Structure, Content, and Scope of Accounting Education, "Future Accounting Education: Preparing for the Expanding Profession," *Issues in Accounting Education* 1 (1990), pp. 168–195.

8. Arthur Andersen & Co., Arthur Young, Coopers & Lybrand, Deloitte Haskins & Sells, Ernst & Whinney, Peat Marwick Main & Co., Price Waterhouse, & Touche Ross [Big Eight], *Perspectives on Education: Capabilities for Success in the Accounting Profession* (New York: authors, 1989).

9. Accounting Education Change Committee [AECC], "Objectives of Education for Accountants: Position Statement Number 1," *Issues in Accounting Education* 5.2 (1990), pp. 307–312.

10. Big Eight, p. 6.

11. AECC, p. 307.

Part 1

Developing a Communication Plan

66 Managers seek information to give their efforts and activities direction. The information they seek is determined by their job responsibilities and their experiences. 99

THE INFORMATION MOSAIC,
SHARON M. McKINNON AND
WILLIAM J. BRUNS, JR.
(HARVARD BUSINESS SCHOOL
PRESS, 1992)

Chapter 2

Analyzing Context and Purpose

J im Smith, the new director of purchasing at DynamiCo, had just finished an oral presentation to a group of line managers. At the end of the presentation, he gave each member of the audience a paper copy of his overhead slides and some financial tables. As he had been taught, he opened the floor to questions. To his surprise, none were asked. After a few minutes, one person said, "You know, Jim, I am at a loss where to begin. I am not even sure what this presentation was supposed to deal with." It seemed to Jim's audience that the *context* and *purpose* of his presentation were unclear. His audience was thinking some or all of the following:

- What was his objective? Why am I here to listen to him?
- How is this relevant to me?
- Does his presentation fit into some larger setting I should be aware of?

To make an effective presentation in the business world, what you want to accomplish and your overall framework for the message must be clear. Many problems with focus and organization stem from not fully understanding context and purpose. In the case of Jim Smith, his audience might conclude that he lacks sufficient knowledge or job skills.

Such an organizational criticism usually points to Jim's context blindness or ignorance. Jim probably has "know-how," but he is not showing it if his audience does not know what his presentation is all about. Thus, Jim seems to be lost and, in turn, has lost his audience. The lost feeling of context ignorance is essentially a communication problem. This chapter includes practical strategies to help you decide where you are relative to the audience and the goals you need to reach.

Assuming you have basic know-how, it's time to think about the contextual *know-when, know-why,* and *know-where.*

In reading this chapter, you will consider a "both sides" alternative to the typical understanding of what constitutes communication, the need to develop a tentative communication plan, the influence of roles in analyzing any communication situation, a comparison between writing and speaking modes in communication, and specific strategies for adapting your goals to the different requirements of written and spoken communication.

Before we discuss specific strategies, let's expand the concept of communication introduced in Chapter 1. Although some may believe that communication is merely the accurate conveying of information or meaning from one person to another, this is a very limited view of the dynamics of communication.

Communication: Beyond Transmission

THE TRADITIONAL MODEL

A noted organizational consultant used to ask business executives in his workshops (before gender neutral language was common in everyday usage) to evaluate the following definition of communication:

> When Person X "communicates" with Person Y, what happens is that he transmits meanings (information, facts, ideas, etc.) from his mind to the mind of Person Y.[1]

In the discussion that followed his purposely inadequate definition, "virtually 100 percent" of the hundreds of executives he asked thought that such a definition was clear, succinct, and effective in defining the responsibilities of the business communication process.[2] After all, they must have reasoned, surely the business environment is one in which some people have the appropriate information and others do not. Thus, those needing information must receive the proper facts and ideas from those who possess information. This *information transmission model* is often taken by executives to be the natural way communication should work. Simplified somewhat in Figure 2–1, communication is understood by such persons as a transmission. However, negative effects from this top-down, one-way

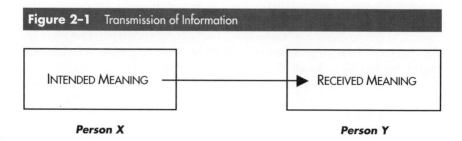

Figure 2–1 Transmission of Information

INTENDED MEANING ⟶ RECEIVED MEANING

Person X *Person Y*

philosophy stimulated recommendations for increased "people skills" in business education, as reported in an influential national research study.[3]

INTERACTIVE COMMUNICATION

The guiding notion of the workshop in which these executives participated improved on the traditional transmission model explained above. Human communication, they discovered, is essentially a mutual experience in which each party in the process potentially influences the others. Thus, a meaning for a message does not start with the sender who deposits it within the receiver. Instead, all communicators are both senders and receivers of messages, often simultaneously. Meaning is uniquely created (in a sense, built or constructed) in the interaction or imagined interaction between the communicators. In addition, the meaning between any two communicators will necessarily differ when compared to any other communicators, even if the content of the communication is exactly the same. If you receive a message regarding some problems with a corporate tax return, your interpretation could be different if it came from a colleague rather than an IRS examiner. This is an *interactive view* of communication that stresses the importance of meanings that the communicators build together:

- In most communication situations, all parties in the process potentially influence the other(s). In most face-to-face settings, the mutual influence is simultaneous.
- Unique meaning emerges in the interaction between the communicators.
- The meaning between any two communicators is different from that between any other communicators, even if the content of the communication is exactly the same.

By *communication* we mean the process by which humans interact with messages to structure and make sense of any situation. Such communication may be verbal and symbolic (like writing/reading or speaking/listening) or nonverbal and behavioral (like interpreting posture, gestures, voice tone, or how much time your boss allots for lunch).

Figure 2–2 shows a diagram of the interactive approach to communication. Although it's slightly more complicated than Figure 2–1, it can help you anticipate a wider range of realistic communication problems. Instead of just a one-way message, Figure 2–2 shows that meaning is based on the interaction of the sender and receiver given the history, skills, and style of each; the relationship between them; the interpretation and feedback each gives; and the context of the message. Central to every communication context is the existence of a relationship between persons. The relationship may be immediate, as with face-to-face interaction, or implied and distant, as with most written texts. Meanings are not directly transferred but interpreted by people according to the history, skill, and style of each person.

As you plan your messages, keep in mind that it is an interactive process: the audience and you ideally will have a common or shared understanding of each

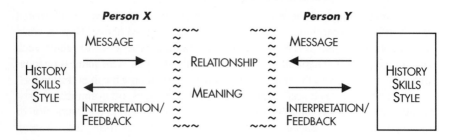

Figure 2-2 Interactive Communication

other's messages, even though all people interpret messages somewhat differently. This has the following implications for organizational communicators:

The both-sides issue. Readers and listeners are just as important in determining the ultimate meaning of your communication as writers or speakers. When we are listeners or readers, we can examine the effectiveness of the message. Using this perspective, you can improve your ability to send messages to others.

For example, if you receive a report on how a certain financial accounting rule is being applied at your firm, beyond just reading the report, you can use this as an opportunity to see what you think is useful and not useful in the communication. How would you have done it differently? What are its strengths? Which of your needs as the intended audience were well taken care of and which were not? You can use this analysis to help you improve your own writing ability as well as to give feedback to the writers of the report.

The relationship issue. A single message you send can have vastly different meanings for different persons, depending on their previous experience, their relation to the source, the communication setting, and many other contextual factors. The message neither carries nor creates meaning; meaning is created out of the relationship between the person(s) and the message(s).

Junior staff accountants are accustomed to receiving messages evaluating their work from senior staff and managers. If they received a similar message directly from a partner, it would probably have a different meaning. Consider how you perceive content and intent of a message if it comes from a superior, a peer, or someone who reports to you. This is part of considering both sides when you are constructing an oral or written message or report. As we so often stress in this book, you must address your audience and task appropriately.

The interpretation issue. When you communicate, your listener or reader may create private inner interpretations to give tentative meaning to your actions. This is a shadow dimension of communication that has little to do with a speaker or writer's intent. Whatever you intend to say, or even if you intend to say literally nothing, others will interpret your words, posture, letterhead, timing, and even your silence or absence.

You may not think anything is being communicated, but on the shadow side of communication, meaning is quietly being built. The silent stare of a CEO or the tone of how an employee is summoned by a supervisor is inevitably interpreted by the employee; that interpretation might or might not match what was intended by that look or tone of voice. How were you summoned in your parents' household? Did different content (e.g., the use of both your first and middle names) or tone imply different meaning to you? Were you always right in your interpretations? Remember that organizational life can be significantly more complicated and prone to misinterpretation than most family interactions.

The creativity issue. Writers and speakers often are not fully aware of what they need to communicate until they encounter an audience. Thus, you might come prepared to discuss some topic but find that it is more important to deal with another, more basic one. The traditional model of communication transmission implies that the writer or speaker already knows what he or she wants to say, and that the audience will, or should, receive that prepackaged message. However, communication is not simply the expression of predetermined meanings; it's the creative process of discovering together with an audience what you, and they, know and can know.

Obviously, you often need to have your intentions understood. Communication planning is a goal-directed activity in which we try to accomplish certain objectives. However, you should not be so concerned with following plans that you inhibit the natural creative potential of communication, which is collaboration with your audience.

The context issue. You can make almost any statement in such a way that negates its clear message. Imagine the first day of your new job, as you dress for success. You ask a friend or your spouse, "How do I look?" The reply is, "Oh . . . you're going to wear that? Interesting choice." Although the dictionary definitions of these words are clear enough to you and seemingly neutral at the literal level, few would fail to recognize the implied criticism. The face value of the total message cannot be separated from its context.

Context awareness is the basis of irony, of friendly kidding, of sarcasm, of subtle suggestion and sly digs, and of social tact as well. If you're context-blind, you may feel complimented when the speaker intended no such thing, or feel put down when he or she thought you would hear positive feedback. Effective communicators interpret as context, among other things, the tone of voice, the statement's placement in the flow of a conversation, the history of communication habits and preferences of your partner, and the immediate physical environment. Without an awareness of how such clues operate, the dictionary meanings of the words will not help communicators much. The importance of context is the reason why it is inadvisable to offer a collection of communication rules and prescriptions to new accounting professionals. Though rules may sometimes help, they do not really teach the learner to be flexible in encountering new situations.

INTERACTIVE NATURE OF COMMUNICATION
··

Communication is an interactive process: The audience and you must have a common understanding of each other's messages.

- **The both-sides issue.** Readers and listeners are just as important in determining the ultimate meaning of your communication as you are as writer or speaker.
- **The relationship issue.** A single message can have vastly different meanings for different persons, depending on their previous experience, their relation to the source, the communication setting, and many other contextual factors.
- **The interpretation issue.** When you communicate, your listener or reader will create private and tentative inner meanings for your actions.
- **The creativity issue.** Writers and speakers often are not aware of what they need to communicate until they encounter an audience. Communication is not simply the expression of predetermined meanings. It's the creative process of discovering together with an audience what you, and they, know and can know.
- **The context issue.** You can make almost any statement in such a way that negates its apparent literal message. Message meanings change as contexts change.

With these basic concepts in mind, let's explore how to plan for effective communication.

Developing a Communication Plan

The first step toward effective communication is to create a communication plan: a tentative design for action. Trying to create a written report or an oral presentation without some sort of plan is like trying to prepare tax forms without knowing the IRS regulations. A communication plan helps you to focus on your goals as a communicator and also serves as a guide and point of reference for you to use as you draft and revise your message. The four main components of a communication plan are:

1. **Purpose.** Establish what you hope to accomplish within a communication situation (your task) and how you can best meet these objectives. For example, if you are dealing with an operational audit of a quality problem in the plant, consider whether you want to lay out information you have discovered, give options for solving the problem, and/or make specific recommendations.

2. **Role analysis.** Define your role within this specific communication situation. A *role* is a well-understood set of expectations for someone's behavior in a context. Your role might be internal (a member of the

organization) or external (someone from outside the organization). Some examples later in the chapter show that your role affects what you want to say and how you want to say it.

3. **Audience analysis.** Determine who will be receiving the message. This means thinking about their goals; their responsibilities; their beliefs, attitudes, and values; and what motivates them to communicate as they do. Chapter 3 deals with this in detail.

4. **Content framework.** Plan how you can frame the message to meet the audience's needs and expectations. Know whether you are presenting an independent message or whether it fits into a larger framework (and how it fits). Assess what you have to say and whether you have researched the topic thoroughly enough to request readers' or listeners' time and careful attention.

Plan how you will organize your information. For example, having an hour to present some ideas about new trends in manufacturing accounting in an informal meeting within your department is a different challenge than having 15 minutes to brief the company president on the same subject. Surely you wouldn't tell the president, "No, thank you; I'd need at least an hour of your time to summarize these complex changes adequately." Instead, you would try to frame the content to fit the task at hand. Each of the components of a communication plan will be considered in detail below and in subsequent chapters.

Creating an effective communication plan may involve developing a detailed outline, making extensive notes, and researching audience background. On the other hand, your planning may have to occur with only a few minutes of thought and only a few notes or reminders.

The complexity of a communication plan and the time needed to create it are determined by the time frame within which you're working and the complexity of the communication itself. Planning a brief memo may take only a few minutes of

Developing a Central Thesis (Purpose and Outcome)

Over the past several months, you have been head of a team looking at the purchasing function at a client company. Over this time, you have interviewed the purchasing personnel, vendors, and those who use the purchased materials and items. Your team has looked at forms, software, hardware, costs, and so forth. It's now time to make a final report to the client company that includes your recommendations.

Task: Write a statement reflecting your objectives (purpose of the report and its value to the client). This statement should be no longer than three sentences.

thought, while planning a detailed analysis of a tax question may involve several weeks of research and planning.

Many busy accountants make the mistake of thinking there is no time to create a communication plan. They want to get right to the report itself and avoid what they see as an unnecessary waste of time. Communication developed without careful planning, though pleasant in some informal situations, often lacks focus or direction in professional contexts. Without a clear plan, you may lose sight of your goals; you may create a message that is useless for your intended audience; or you may leave out important ideas and information. Careful planning allows you to think through the entire context so that, unlike Jim in the opening example of this chapter, your message and purpose will be clear to your audience.

Communicator Roles

The four components of a communication plan (purpose, role analysis, audience analysis, and content framework) are interrelated. While we discuss aspects of them separately, do not think of them as if they were discrete. Let's start with role. Here, *role* means those behaviors and communication habits that audiences and communicators come to expect of a particular person in a particular context.

People might think of roles as being played by actors on a stage, in movies, or on television, with scripted and predetermined interactions among the players. However, in most social situations, people assume roles and anticipate each other's actions. Normally, a person's objective in a given situation determines to some extent his or her role.

For example, if you want a group to change its policies, you may adopt a leadership role in order to persuade rather than taking on a role of passive observer simply to acquiesce to other members' wishes. You might assume the role of teacher when you help others obtain information. Other common roles are facilitator, catalyst, mediator, negotiator, supervisor, confidant, parent, and so forth. Each role has customary and expected behaviors associated with it; leaders, for instance, are expected (in Western culture) to be relatively active, decisive, and talkative, while mediators are expected to be unbiased and to use neutral language. However, knowing your role also involves some subtle elements.

- Roles and the implicit rules that accompany them do not prescribe exactly what should happen in an interaction, but they do tend to enclose people within some boundaries for describing behavior that is socially acceptable.

- You need to be both *role-sensitive* and *role-flexible* at the same time to be an effective communicator. Be sensitive to the role implications of a given situation, and flexible with your role to meet the expectations of listeners or readers.

- You must meet enough of the role expectations of your audience so that they will perceive you as credible (expert and trustworthy in the given task), but you can't be locked into a straitjacket of expectations that eliminate your personal style.

ACCOUNTANTS' ROLES

As an accountant, you are expected to perform many specific roles, including statistician, decision maker, investigator, adviser, observer, and questioner. Your technical roles within the organization will often be quite specialized and, to some extent, arbitrarily chosen by superiors. Yet your responsibilities will create an impact far beyond their obvious influence. For instance, you might be on a team of accountants that has analyzed a company's controls for handling cash, making deposits, creating purchase orders, and similar functions. As either internal or external auditors, such a team might be asked to just provide a neutral, objective, and scientific report that outlines facts but makes no recommendations. Alternatively, the team could act as advocates and recommend a course of action they believe justified by the data. If you are part of the team doing the annual independent audit, you would be expected to provide management with specific recommendations and rationale. Your team can explain, as teachers might, or lead and persuade, as salespersons might. The differences, as you will see in later chapters, are crucial. What gives you credibility as an objective researcher may not increase your credibility in an advocacy role. The messages you will send to your audience will be necessarily different.

In this situation, it is also important to consider how your team's roles are related to, or defined by, others. Will an internal team be seen by the board as independent or as mouthpieces for the executive who assigned them the task in the first place?

ROLES IN AN ORGANIZATIONAL CONTEXT

An accountant's role makes a significant difference in how a report should be written and circulated in the first place. Roles make rules, and rules make or break people in organizations. Here are seven practical guidelines for making role decisions in an appropriate context within organizations:

1. **Analyze the organizational culture.** Examine the norms, the ways of talking, the customary topics, the recurring stories, and the dramatic tensions of the group. Some would call this an organizational or *corporate culture,* by analogy to the differences between ethnic and national cultures. What will work in a Japanese company will not be effective at the Kansas City Brass Works. Additionally, Kansas City Brass Works and its culture cannot be directly translated to Mutual of Omaha's corporate culture. A joke

---

Thesis Explored Again

In the previous Practice exercise, role was not defined. You had to decide what the role should be.

Task: Write two different thesis statements for the prior exercise based on a role of (1) neutrality and (2) advocacy.

PRACTICE

heard as witty in one office may be considered gauche and tasteless in another.

2. **Analyze the physical context.** Think about where you will be communicating your ideas and listening to others. It might be in an auditorium with audiovisual facilities or in a small conference room with a circular table. All other things being equal, an auditorium invites a more formal presentational style, while a small conference room encourages feedback, discussion, and interchange of ideas among audience members. If you are insensitive to the constraints of the physical environment, your message may be lost in the shuffle.

    Physical context can also have an impact on written communication. Simply ask yourself such questions as: Where will my report likely be read? What else is likely to be distracting the reader while he or she considers my ideas, and how can I take these distractions into account? For example, in one office, workers noticed that a manager carried her morning mail around with her and read it while she made the rounds to visit different department heads. This should affect the complexity and elaboration of the reasoning you use in a memo to her. Further, you might want to suggest a follow-up appointment with her to explain key passages.

3. **Analyze your audience's investment.** In other words, determine who (within the context in which you work) has a stake in what outcome. In our previous analysis of the example of internal controls, people at different levels of management will be personally involved in how cash is handled or how purchase orders are processed. No matter how intelligent and objective a report might be, it will always be read through the subjective filters of such personal involvement. Who is a champion of the project and who does it affect? Whose image could it enhance and whose might it question? Unfortunately, such role complications are often not anticipated and the chances for effective communication are diminished.

4. **Analyze audience characteristics.** This is a small-scale version of how survey researchers discover the relevant social and demographic characteristics of a group of consumers or voters. A demographic profile is not foolproof, but it goes a long way toward helping communicators predict the issues, styles, roles, and strategies that would be most appropriate for a given audience. When an advertising agency buys commercial time on "Frasier," it has access to the network's demographic research. It knows the age, gender, racial makeup, mean income, purchasing habits, and average impulse-spending habits of the viewership. It's no accident that the resulting commercial appeals in a consistent way to the values suggested by such consumption groups. If you ultimately find the commercial irritating, you might be tempted to call the whole process manipulative, but advertisers simply defend it as a coordinated system of communication in which an audience as a whole is more likely to gain access to the information and products it wants. In the same way, it is not inherently unethical or

manipulative for you to provide information that is well adapted to audience needs, habits, listening styles, and goals. Chapter 3 discusses audience analysis and focuses specifically on this crucial communication problem.

5. **Analyze your personal investment.** You might have strong feelings about or feel ego-involved in the ideas you will be explaining, or you might be unable to hear or unwilling to consider alternative interpretations and disagreements with your point of view. Do you tend to see suggestions, objections, and modifications as confrontations, as nitpicking, or as subversions of your good efforts? Unfortunately, many communicators unwittingly take revision as rejection. A healthier perspective is to understand that your message is probably not itself a final product but a contribution toward a productive exchange of ideas.

6. **Analyze your personal strengths.** Consider whether you have been more successful in writing situations or in speaking situations. Look at whether you are more successful with an expansive expository writing style or with a more concise summary style. You may have had different kinds of success with formal public speaking or informal conversational dialogue. If necessary, interview friends and co-workers you trust to provide you with reliable and honest feedback. If you have a choice in the form, format, or context of your communication, you can use this assessment as a basis for your choices. Of course, books such as this one seek to expand your repertoire so that ideally you'll be flexible enough to communicate in a style appropriate to any given audience or context.

7. **Analyze your communication plan from an audience perspective.** Understanding communication as an interactive process means that you see a speaker's or a writer's power to influence audiences, but you also recognize the power of the audience. It's especially important, given this interdependence, to decide what you want to *accomplish with* (not *do to*) the audience. You have to decide if you want to inform the audience, persuade them to alter their attitudes, coordinate your goals and needs with theirs, motivate them, entertain them, or some focused combination of the above.

PRACTICAL GUIDELINES FOR MAKING ROLE DECISIONS

- Analyze the organization's culture.
- Analyze the physical context.
- Analyze your audience's investment.
- Analyze audience characteristics.
- Analyze your personal investment.
- Analyze your personal strengths.
- Analyze your communication plan from an audience perspective.

PRACTICE

Role-Audience Pairs (Dropping Cosmetics)

The president of a chain of pharmacies wants a recommendation (in memo form) regarding whether the company should drop the sale of cosmetics.

Task 1: What information would you want before writing your memo?

Task 2: How would your memorandum differ given the following role-audience pairs?

Role (You)	Audience
Treasurer of the company	President of the company
Outside consultant	President of the company
Outside consultant	Treasurer of the company
Staff accountant	President of the company

Differences between Written and Oral Communication

One of the most basic differences between oral and written communication is immediacy. As a writer, you face the dilemma of being removed in time and space from your communication partners. Therefore, contextual analysis and audience analysis may be less precise. Writers can shape their message for a given audience, but usually cannot control whether the readership is limited to that audience. Even personal letters or e-mail messages, intended for a specific and private reading, will often be studied by unintended eyes, as subjects of unauthorized biographies or celebrities in the news recently have discovered the hard way. Additionally, writers understand that readers, unlike listeners, cannot contribute to the immediate development and modification of the original message. Feedback is deferred if not discouraged.

Therefore, writing becomes more like a monologue, and historically much more dependent upon the authority (literally, the "author-ness") of the writer. (In the organizations where you work, does the staff at any level write as many memos as the supervisors? When the IRS contacts you concerning your tax return, do they phone you or drop by the house?) Readers, in a sense, are dependent upon the directions of writers, at least if they want to remain in the communication relationship. However, readers control the context in that they may choose either to continue or to stop reading.

By contrast, speech is more interactive because any utterance or nonverbal clue potentially invites immediate responses or spontaneous requests for clarification. Even in a formal public speech, most speakers are available for some forms of interruptions and often explicitly structure question-and-answer periods. Although both readers and listeners are participants in communication, listeners are more overt participants.

PRACTICAL CONSIDERATIONS

What does all this mean in practical terms? Think about power relationships. Among other things, in most organizational contexts, writing a message (a report, a memorandum, a letter, an e-mail message, a note) is itself a small expression of power regardless of its content. Writing, especially formal writing, presumes a degree of organizational permanence and further presumes that the message might be disseminated. While things like e-mail are much more informal, in all cases the simple act of writing something asserts that the writer is confident enough to subject the message to response. In one sense, writing pins down the writer to a particular content, but it also pins down the reader to the writer's authority to direct or mandate a basically one-sided situation.

As a writer, therefore, you must be prepared to have your work scrutinized and analyzed, especially by those who might disagree. As the "letters to the editor" pages in newspapers demonstrate, readers are much less tolerant of errors than are conversation partners, and often more incensed at divergent opinions. In contrast, an accepted principle of conversation is that listeners should try to be cooperative with a speaker's ideas and not "rock the boat" unless absolutely necessary.

On the other hand, speakers might want to anticipate the impermanent nature of talk. While readers can literally *refer* back to what they read yesterday if they want to check on detail, listeners can only *try to recall* what they heard yesterday. They usually do so imperfectly. Some listeners rapidly forget (or never register) both content and context. Listeners, after a time, may dissociate content from context, forgetting, for example, where they heard that the Apex Division is being shut down and all the reasons for its closing.

Face-to-face speech, unless it is captured electronically, trades off permanence for personalization. This is why many managers talk an idea around or send up verbal trial balloons before committing an idea to paper; they want the personalization along with the permanence they will eventually get with a directive or memo. It's also why effective public speakers like Maya Angelou, Jesse Jackson, or Elizabeth Dole build into their speeches memorable phrases, alliteration, and striking metaphors; these memory aids give the message a degree of permanence along with the personalization of seemingly spontaneous speech.

Understanding basic differences between writing and speaking is crucial for improving your communication initiatives. With these differences as backdrop, the next two sections present specific strategies for coping with some of the contextual demands of written and oral communication.

Strategies for Written Communication

Our culture seems to have a reverence for the written word. How many times have you thrown away a book, even if it's an out-of-date accounting text? Most people cannot bring themselves to toss old books; they may sell them at a garage sale or donate them to a library, but they do not throw them away. Printed words have a

---

PLAN AHEAD OF TIME

- Prioritize your objective(s).
- Decide what format will best accomplish your objective(s).
- Analyze the context that has prompted your writing.
- Analyze the context in which your text will be read.

---

value for us on some unconscious level. When you write for a class or in a professional context, you have to deal with this "set in stone" quality of writing. One way to feel more comfortable with the seeming permanence of your writing is to prepare ahead of time by planning carefully and analyzing context as it relates to writing. Remember to:

1. **Prioritize your objective(s).** Decide on your primary and secondary goals. Are you providing information (secondary goal) in order to persuade (primary goal)? Are you providing an analysis for a new client (secondary goal) in hopes of building a long-term business relationship (primary goal)? Until you are clear on what you hope to accomplish, you are probably not ready to begin sharing your writing. Your choice of words, the organization of your text, even the placement of the words on the page are all affected by the priority of your objectives. While you might decide to reorder and/or reword these objectives as you write, refer back to your objectives on a regular basis to ensure you are staying on track.

2. **Decide what format will best accomplish your objective(s).** Can you best meet your goals through a letter, a brief memo, or a technical report? If you have developed a simple solution for an auditing concern of a client, choose a brief letter rather than a 15-page detailed report. Your objectives ought to shape your choice of written format. Further, how your writing is packaged can also be a kind of message to readers. Your writing, whatever its internal style or language, can be made more or less formal by such choices as what kind of folder or envelope in which to enclose it, how it is delivered, and whether it includes a personalized handwritten greeting. In addition, decide what content should be presented in paragraph form, what in lists, and what in tables, charts, or graphs. This is important as well for more informal forms of communication. You might also have the choice of using e-mail.

3. **Analyze the context that has prompted your writing.** Are you responding to a request for information or analysis? Are you initiating the communication? Is your text to serve as a follow-up to a previous communication situation? (If so, have you included enough previous context to remind readers of the broader picture of what is at stake?) Are you working as an employee of an organization (internal role), or are you writing as a consultant (external role)? You may need to make explicit reference to the appropriate context.

4. **Analyze the context in which your text will be read.** Will the text be complete in itself, or is it to be part of an ongoing process? Has your audience requested that you write this text? Will your reader be pleased with what you say, or will your text be a stimulus for conflict? Remember that since you probably will not be present when your text is read, you will not have the opportunity to clear up misunderstandings or smooth ruffled feathers. You need to envision how your text will be received and plan your message accordingly.

Strategies for Oral Communication

Most of the factors above also apply to oral communication, such as prioritizing your objectives, deciding what format will best accomplish your objectives, and analyzing the context. However, speakers must consider additional factors. In speaking situations, you will be seen and heard in an immediate context; that is, your audience will evaluate your credibility as a direct part of your message. Who you are, how you conduct yourself, and how you appear to your audience will to a large extent determine their involvement with your ideas. You are part of your message. This fact is crucial, and its importance is impossible to overstate. Therefore, remember:

1. **Dress appropriately and professionally, taking your cues from the organizational culture and the habits of the people to whom you'll speak.** When in doubt, choose suits and professional wear that are relatively formal; men and women can often remove a jacket to become less formal if the situation warrants. However, if you arrive dressed informally, there is no way to formalize your attire on the spot.

ORAL PRESENTATIONS

- Dress appropriately and professionally, taking your cues from the organizational culture and the habits of the people to whom you'll speak.
- Make sure ahead of time that you have a clear and realistic idea of how long you're expected to speak to and/or interact with an audience.
- If possible, visit the site of your speech or conference beforehand.
- Remember that you are part of the context for the audience.
- In your introduction, refer briefly and preferably in good humor to any obvious features of context, such as excessive room heat or lateness of the meeting.
- Introduce your purpose and yourself.
- Decide if your context and purpose demand precise, exact wording.

2. **Make sure ahead of time that you have a clear and realistic idea of how long you're expected to speak to and/or interact with an audience.** Analyzing context will help you plan the appropriate degree of detail for your talk. Although some people might fear droning on to a bored or restless audience, many business speakers also err by being less thorough and detailed than an audience has a right to expect. If an hour is allotted to considering your report, an initial seven-minute presentation probably will not lay the foundation for subsequent dialogue.

   In addition, consider your presentation in light of what others will be presenting before and after you. If you have an allotted time, make sure to stick with it. Few things anger audiences more than a speaker intentionally ignoring a time limit. It's especially bad form to say something like this: "I know I am out of time, but I want to make the following points." This implies that what you have to say is more important than what subsequent speakers will say and more important than other things that your audience is scheduled to do when you are finished.

3. **If possible, visit the site of your speech or conference beforehand.** Get the lay of the land. Decide tentatively where you'll sit or stand and where you'll place any materials to assist you. Arrive early, if appropriate, and be fully prepared before the audience gathers. If you cannot visit the site ahead of time, at least inquire about the size of the room and the expected size of the audience.

   Become familiar with any equipment or technology you might be using. Too often speakers do not anticipate local problems with setting up an Internet connection or switching from PowerPoint slides to a video clip. Becoming familiar with the equipment also means knowing what you will do if it fails. For example, if you plan to use a laptop computer with a color projector and PowerPoint slides, have an alternate plan if some part of the equipment fails and cannot be fixed. Such a contingency plan might mean carrying a backup set of color transparencies to use with an overhead projector, or having handouts of your slides in case the equipment malfunctions.

4. **Remember that you are part of the context for the audience.** Your mood and even your willingness to smile and chat before the event will set the stage for smoother communication later. Moods and personalities are often subtly infectious. If you look nervous and disinterested, the anxiety level of your communication partners will increase. You may notice your audience's anxiety when you start speaking, and, in a negative vicious cycle, you might become concerned as you speak that your audience does not like you. It isn't necessarily that they don't like you; more likely, they are mirroring your own attitude.

5. **In your introduction, refer briefly and preferably in good humor to any obvious features of context, such as excessive room heat or lateness of the meeting.** Your audience wants to know that you are there with them,

physically and psychologically. However, do not dwell on problems, reminding the audience of other things to consider aside from your main purpose for speaking with them.

6. **Introduce your purpose and yourself.** The introduction should clarify your perspective (1) relative to the audience, (2) relative to the organizational problem you're attempting to address, and (3) relative to the audience's need to hear what you have to say. Establish credibility succinctly by including whatever personal credentials are relevant, without boasting, and by describing which systematic methods you employed in coming to your conclusions. (By the way, lengthy prepared jokes in introductions, unless you are near-professional in adapting them to the immediate audience and context, usually fall flat.)

7. **Decide if your context and purpose demand precise, exact wording.** If so, you may decide to read some portion of your message directly from a manuscript, as a head of state might do to introduce a news conference. Sometimes, the wording of a statement must be so diplomatically correct in a sensitive situation that speakers wouldn't trust spontaneous styles of talk. These situations are rare. If you encounter one, practice reading the relevant passage aloud often, rehearsing the inflections that fit your intended audience.

Most of the time, plan to speak extemporaneously with appropriate visual aids and/or handouts. (See Chapter 6.) *Extemporaneous speaking* means that you are speaking spontaneously, but have prepared your communication plan carefully, and are well aware of the content and context. You have an outline of topics, but are speaking conversationally in

Role-Audience Pairs (Budgeting)

You have been asked to explain the concept of *budgeting* and to give some examples.

Task 1: How would the goals, message formats, and contexts of your presentation differ given the following role-audience pairs?

Role (You)	Audience
Staff accountant	Operating manufacturing managers
Staff accountant	Hourly workers on the production line
Staff accountant	Secretaries
Teacher	High school students
Staff accountant	High school students

Task 2: How would a written presentation differ in each of the above pairs from an oral presentation?

adapting them to the immediate situation and audience. You must even be willing to change your outline if an unanticipated opportunity or question emerges. Extemporaneous speaking differs from both *impromptu speaking* (speaking off-the-cuff, winging it) and *manuscript reading* or *memorized speaking* (which come across as very formal and inflexible).

These last three items can also support written communication. There might be a specific context problem to mention (e.g., the need to compress a good deal of information into a required two-page limit). As we discuss in later chapters, an introduction sets the stage for what follows. Finally, the same admonition about precise, exact wording applies to written communication as well.

Summary

Communication is often taken for granted as basically a message transfer process. In emphasizing the issues of context and purpose, this chapter has taken a different perspective. In order to be an effective communicator, you must realize that readers and listeners are ultimately as much in control of the process as are writers and speakers.

Therefore, this chapter highlighted five general issues. First, communication was defined as the process by which humans interact with messages to structure and make sense of any situation they share in common. Second, we stressed the need for a communication plan, especially in organizations where role differences are clearly defined. Third, we examined the influence of communicator roles on planning organizational messages, whether in writing or speech. Fourth, we compared written and oral communication to see which differences might be critical for communicators to consider. Finally, we surveyed the two modes, written and oral, to emphasize strategies particularly appropriate to each.

Your next step as a communication planner is a more careful consideration of who will be reading or hearing your ideas. Audience analysis is a crucial component of communication success.

How to Approach Practice Tasks

DEVELOPING A CENTRAL THESIS (PURPOSE AND OUTCOME)

This is the first of two exercises where you are asked to state the purpose of a report. In this one, role is not defined; it is something that you have to assume. Note how hard it would be to state the purpose of the report without at least making an implicit assumption regarding your role. You need a perspective from which to pose your central thesis. Your two- or three-sentence statement of purpose should incorporate context. This includes items such as the following:

- Will this report stand alone or be part of some ongoing process (e.g., is a preliminary report to be followed in a month with a final report)?

- What is your purpose: to inform, persuade, provide alternatives for consideration?

- Most important, what do you want to accomplish? A statement of purpose should leave the listener or reader with a sense of why what follows adds value.

THESIS EXPLORED AGAIN

Now your role is specified. Let's look at a sample of what you might say in each of these roles.

- **Neutral role:** This report is the culmination of three months' work by our team to analyze purchasing at Fantasia Corporation. In it, we present what we have found and provide a balanced set of perspectives on relevant issues and questions. As a result of this analysis, management will be able to assess possible actions.

- **Advocacy role:** This report is the culmination of three months' work by our team to analyze purchasing at Fantasia Corporation. In it, we present what we have found and provide a set of recommendations with accompanying rationale. As a result of this analysis and set of recommendations, you will be able to improve your purchasing function and save over $100,000 each year.

ROLE-AUDIENCE PAIRS (DROPPING COSMETICS)

These tasks cast you in different roles writing to different audiences. To start, no matter what the role, the decision of whether to stop the sale of cosmetics should be placed in the context of the overall direction of the company. Thus, in responding to the first task, your decision about what information to include is affected by the purpose of your analysis. The effect of dropping this segment of the business can be looked at from both a strategic perspective as well as a short-term analysis. Not only would you want information relating to the annual lost revenues and costs saved, but you would also need to know all the interdependencies between selling cosmetics and selling other items. Depending on these relationships, you would want not only quantitative financial information (e.g., labor savings, sale of fixtures) but also nonfinancial quantitative information (e.g., number of people who buy cosmetics who also buy greeting cards) and qualitative information (e.g., how customers feel about this decision).

In Task 2, each of the role-audience pairs affects how you would present information. The general format of the presentation (a memo) is already defined. Let us compare the difference between being a staff accountant writing to the president and an outside consultant writing to the president. What does the president expect of a staff accountant as compared to an outside consultant? How can you (in either role) advance your standing not only with the president but also with the treasurer? What is the organizational culture regarding these different roles?

As a consultant, you would probably have a relatively clear charge—an understanding of the task to be accomplished by your engagement. As a staff member, this understanding might be much more informal. How your role will affect the strength of what you recommend is an open question. However, in all the role-audience pairs, you would define the alternatives open to the company, the relevant information to consider, what you recommend, and what you see as the strengths and weaknesses of your recommendation. In developing this general framework, what is the personal involvement of the president, treasurer, or others in this decision? Perhaps the president thinks that you really cannot have a pharmacy without the sale of cosmetics; the president is asking for this information in response to a query from a major stockholder. Knowing this might allow you to shape your recommendation in a way that does not offend either party. This is different than a situation where nobody has any personal feelings about a purely business question.

ROLE-AUDIENCE PAIRS (BUDGETING)

In Task 1, consider different audience needs. Probably none of your audience groups will be proficient in accounting jargon or know much about how budgets can be tied to projected (pro forma) financial statements. Let's assume that management wants operating manufacturing managers, hourly workers on the production line, and secretaries all to have some knowledge of budgeting. Awareness of budgets (and what things cost) allows people to make rational choices (e.g., use of overtime by a supervisor, better preventative maintenance by line workers, or quantity purchasing of supplies by the secretarial support staff).

It's important to use plain English. You would use language and examples that are familiar to your audience. For hourly workers, you might relate things to a family budget, where some items have to be paid every month like the mortgage or rent, and basic utility bills. Food would be at some minimum level. From there, you can go into discretionary use of whatever is left over, such as saving for a future purpose like making a down payment on a new automobile. This can now be related to an illustration within the workers' experience on the job and sphere of operations.

Note how your task changes as you talk to high school students. Here you might be part of a career day as a staff accountant (perhaps an alumnus of the high school) and want to talk about budgeting as an example of what accountants do or what is done in a more general business setting. If you were the teacher of the class, your objectives might be broader and might lead to some problems that students would use to practice looking at the concepts you were going to explain.

Task 2 asks you to consider what you would do differently if you delivered your report orally or in writing. In either case, you want your message to be understood by and accessible to your audience. In an oral report, you might choose different clip art or cartoons depending on your audience. In a written report, consider what you would want to put in paragraph form and what you would want to put as a chart or other form of graphic illustration. In an oral report, consider that your audience will use their eyes and ears to gain understanding of your message. While you might leave them with a handout, their general impression is going to be received in real time as you make your presentation. Written communication allows reflection and later reference.

Endnotes

1. W. C. Redding, "Human Communication Behavior in Complex Organizations: Some Fallacies Revisited," in C. Larson and F. E. X. Dance (Eds.), *Perspectives on Communication* (Milwaukee, WI: Speech Communication Center, University of Wisconsin, 1968), pp. 99–112.

2. Redding, p. 101.

3. L. Porter and L. McKibbon, *Management Education and Development: Drift or Thrust into the 21st Century* (New York: McGraw-Hill, 1988).

Chapter 3

Analyzing the Audience

"... AND SO ... AS YOU CAN SEE, MY THEOREM PROVES CONCLUSIVELY THAT I AM TALKING WAY OVER YOUR HEAD."

armen Cepero is a mid-level production manager at Vencedora Manufacturing. She is in the front row for a consultant's presentation on how to manage costs. As the presentation progresses, she finds that she does not understand the terms being used. She hears terms like "structural cost drivers," "executional cost drivers," and "fundamental cost drivers" with no attempt by the presenter to define them or to use plain English instead of relying on jargon. She feels both frustrated and confused.

Tom Kennedy is part of a financial planning workshop run by Linda Pilcher, a CPA and CFA. Pilcher has been talking for about 15 minutes nonstop and has been telling the audience things like: "You know that if you sell stock and get more for it than you paid, you are required to pay taxes on the gain." Kennedy, along with the rest of the audience, is getting restless and bored. Most of what Pilcher tells them is common knowledge for this audience.

Have you ever been in a similar situation where the speaker is either talking way over your head or seems to be talking down to you? This can also happen in reading a memorandum or report. In these cases, your needs are not being addressed. Common sense dictates that all messages should be shaped, at least in part, by the needs of the audience. In this chapter you will consider the importance of audience analysis for successful communication, the difference between writer-centered and audience-centered texts, questions you can ask when analyzing an audience, and audience analysis strategies for both written and oral communication.

The Role of Audience

Almost every day you can see examples where barriers to communication are created by speakers or writers who do not adequately address audience needs. Assume you are in the accounting department of a major firm and have done some tax research. The results are in and your supervisor asks you to share them. Consider the different needs of the following audiences: If you are going to speak to the board of directors or to a broad range of upper-level management, how would this differ from addressing the CFO, controller, and other accounting department managers? What about sharing this information with the tax partner and manager in charge of the company's account at the firm that does the audit? These three audiences will need a different level of details, different language, and different examples. Upper-level management and the board of directors are sophisticated businesspeople. On the other hand, they do not know tax jargon or the intricacies of the tax code. They want a broad overview of the situation with recommendations they can understand and summary support material showing the effect on annual taxes. At the other extreme, the tax partner and manager at the company's public accounting firm are well versed in both jargon and detail. Here, you might resort to shorthand notes, provide many pages of detail for analysis once you have left, and feel free to use as much technical language as you choose. You have the same message to share. How you share it with different audience groups can make a difference in its effect and on how others judge your performance.

Inadequate audience awareness is apparent in many failed and ineffective written communication attempts, as well. You may have had the experience of buying a printer for your computer and being totally frustrated when you tried to set it up, following the directions in the owner's manual. Computer and printer manuals are notoriously difficult for the nontechnical person to use. Most manuals are written on a highly technical level, even though the majority of users are not "computer jocks" and are not conversant with computer jargon.

Just as you may have been frustrated by messages you think are unclear, the people with whom you communicate may also feel a sense of frustration if your words don't fit their experience. Therefore, audience awareness and analysis must be a major component in every communication plan you develop. Your success as a business communicator hinges on your ability to anticipate and meet the needs of an audience. Indeed, the audience must become the focal point of your entire creative/decision-making process. Your audience will affect the format you select, the organizational pattern you use, even the words you choose.

Writer-Centered versus Audience-Centered Texts

To understand why your audience is such a crucial factor in the success or failure of your messages, consider the difference between *writer-centered* and *audience-centered* texts. Sometimes as communicators, we need to address only ourselves. Journals, diaries, and classroom or meeting notes are examples of writer-centered texts. No one else will be using this information but you. You can write in shorthand or even hieroglyphics; it does not matter, as long as *you* can understand the message. Writer-centered texts do not require any concern for audience; you know what you meant to say, so there is little chance for the message to be misunderstood.

Obviously, only a small portion of the communication you will engage in as an accountant is writer-centered. Auditors' tick marks are an example, as are notes you might take when meeting with a client. The majority of the time, you are creating and sending messages meant to be audience-centered and interpreted by someone else.

AUDIENCE AND COMMUNICATION

- Effective communicators never lose track of their audience.
- You need to keep your audience foremost in your mind as you plan and execute your message.

Looking at a Writer-Centered Text

PRACTICE

Locate one of your own writer-centered texts from class notes, diaries, or meeting notes. Designate a hypothetical audience for this text (a classmate, a colleague, a professor), and rewrite the information to be audience-centered. What are the specific changes you needed to make in order to accommodate your intended audience? Note how the appearance of the text changes when you are no longer only writing for yourself.

The Process of Audience Analysis

Given the importance of audience and audience-centered communication, you will need to analyze your audience. The following checklist of questions should help you begin the process of audience analysis:

1. Who will be receiving this message?
2. What is my relationship to this audience?
3. What does my audience already know about this subject?
4. What does my audience need to know about this subject?
5. What are my audience's feelings about this subject?
6. What cultural differences might influence how my audience will read or hear messages?

WHO WILL BE RECEIVING THIS MESSAGE?

This question is so basic that it may seem completely superfluous. However, when you anticipate the likely receivers of your message, you often discover an unexpected web of complexity.

Single-person audience. You may know for certain that you will have a single reader or listener. Perhaps you are writing a letter detailing new federal tax guidelines to a specific individual, or you are preparing an oral presentation on tax changes for a one-on-one meeting. You can assume that no one else will be reading the letter or attending the meeting. Your job is somewhat easier when you have a single-person audience; you can make decisions with more certainty because the number of audience factors is limited. Unfortunately, organizational messages are rarely this clear cut.

Multiple audience. Sometimes, you will find yourself preparing to address what we will call a multiple audience. The letter you are writing may be addressed to one person, but you (or the recipient) will be sending copies to several other people as well, or your oral presentation may have been requested by one person, but you will be speaking in front of an entire department. Multiple audiences create multiple problems for communicators. The audience may be homogeneous (persons who are similar in age, level of education, and occupation), but more likely, a multiple audience will represent a variety of backgrounds.

For multiple audiences you may want to design your message for an average audience member. For example, your oral presentation on tax changes may be attended by people who are familiar with federal tax laws as well as by those who do not have a background in taxation. You can shape your overall report for someone with a mid-range knowledge of taxation. In order to include those who are less knowledgeable, you can stop and say, "I realize this information may be new to some of you. I will try to include the definitions you need. Let me know if you have questions or if I'm going too fast." To acknowledge the more advanced awareness of some audience members, you can say, "Many of you are familiar with these procedures, but I felt a brief summary would focus the issues for all of us."

In an organizational setting, the audience might consist of different levels of management. This can present different problems in what you want to say. Perhaps you are presenting a comparison of actual performance against budget. If there are segments of the company that have not been performing up to bench-marked standards and your audience includes an array of that segment's managers (upper-, mid-, and lower-level), what you say and how you say it might be different than if you were just addressing upper-level (or lower-level) managers. We discuss flexibility more in the following sections.

Unknown audience. Often, you will need to communicate with an unknown or unspecifiable audience, one you cannot anticipate precisely even with a demographic analysis. You are preparing a companywide memo, but you do not know who will read it; you are writing a magazine article on auditing software; or you are preparing a speech for a national accounting convention. In each of these situations you cannot be certain of the characteristics of your audience. You will have to make a choice whether to aim for the middle ground (perhaps acknowledging this fact early, but making every effort to include individuals on both sides of the middle) or whether to address those closer to the higher end of the spectrum of what the audience could know. For some writing contexts, you have the option of adding an appendix or list of references for those who need additional background information or for those who want to learn more about your subject. In a speaking setting, what you send out in advance or leave as speaker's notes can accomplish this same task.

Target audience. Many speakers and writers know their ideas will be received by some people in the audience who ultimately have no influence on how their overall message will be used. A speech that irritates or even alienates most listeners, but which impresses the comptroller (your target audience), can still be considered successful. Thus, determine who your target audience is. This might be your entire audience or some specified subset.

WHAT IS MY RELATIONSHIP TO THIS AUDIENCE?

Understanding your relationship to the audience presumes an analysis of the communication context, as discussed in Chapter 2. First consider your position relative to your intended audience. You might be writing or speaking to a peer or a colleague, a professor or your supervisor, a younger classmate or a subordinate. Most of us find ourselves functioning within several hierarchies of relations every day. Within an organization you may be part of a team of auditors; you may have a manager who oversees your work; and you may have an assistant whose work you supervise. Obviously your interaction with these various individuals will be affected and shaped by your relationships with them. These relationships will affect tone (formal or informal) as well as word choice.

In addition to thinking in terms of a hierarchical relationship, you also need to consider the qualities that characterize your relationship to your audience. You could be friends, adversaries, or strangers. The audience might be clients or persons you are hoping to do business with in the future. For example, you will probably have established credibility with current clients; they know you and you have an ongoing relationship. Prospective clients, on the other hand, do not know you. Thus, with the latter audience, you might need to be more formal and to say things in a way that helps to establish you as a credible source. You might state, "The problems you pose seem quite similar to ones we helped address for the XYZ Company." Thus, relationships will be important criteria as you make writing and editing decisions.

WHAT DOES MY AUDIENCE ALREADY KNOW ABOUT THIS SUBJECT?

Answering this question will help you decide how much background information you need to include or how technical your language and approach can be. If you are dealing with an audience that shares your level of knowledge and expertise, you do not need to elaborate on the information you share or define familiar terms. Many readers or listeners can detect condescension and have little time to wade through information they already have. If you are an outside consultant and you are addressing managers of the firm that hired you, you would not want to begin your analysis with a long history of the company. You assume readers or listeners are already aware of the company's past.

In speaking you can rapidly adjust to your audience's level of understanding. You might come prepared to explain the qualities of two different machines only to find that your audience is already familiar with them and does not need this information. Speakers should be flexible enough to adjust to such situations. That is one of the wonderful (and scary) parts of speaking versus writing: real-time and instantaneous feedback.

Overestimating the level of knowledge and expertise of your audience is too common, as readers of computer manuals will attest. Your message will not be understood if you talk over the heads of your audience or use *jargon* (specialized terminology that only experts in a given field are likely to know) that they do not understand. Imagine you are working for a client who owns a small business but who has little knowledge of cost accounting techniques. This client wants predictions about possible profits for the coming year; however, if you send a letter that includes unexplained terms such as *contribution margin* or *cost-volume-profit analysis,* the information will be of little value to your client. If you need to include such terminology, then you will want to define and explain any and all terms that may be foreign to your audience.

Remember that effective definitions are succinct, accurate, and relate the defined term directly to what the audience *already* understands. Thus, concrete examples are more helpful than textbook precision. For example, in a talk you might briefly observe: "*Overhead* refers to business expenses that are not directly included in the merchandise we sell—such as the rent we pay on those fancy new copy machines."

Anticipating the average level of audience knowledge accurately will save your audience from being bored or being insulted. During an oral presentation, you should be aware of when to cut back on what you had planned to present or

Rewriting Based on What Your Audience Knows

C onsider the following: "Product costs involve inventoriable costs as compared to period costs, which are expensed in the period incurred. Besides direct and indirect variable costs, common manufacturing overhead is applied to each unit of product using appropriate drivers."

Task: Rewrite this message for each of the following audiences attending separate educational seminars on manufacturing accounting. (*Note:* You might have to look up the technical definitions yourself, to ensure your translations will be both accurate and helpful.)

1. A group of production managers.
2. A group of sales managers.
3. A group of union officers who represent hourly production workers.
4. High school students.

when to fill in the blanks to bring your audience up to speed. In Chapters 6 and 8 we discuss more about how to discover what your audience already knows.

WHAT DOES MY AUDIENCE NEED TO KNOW ABOUT THIS SUBJECT?

This question builds on answers to the previous question. Suppose you are preparing an oral presentation for a client in response to an auditing procedures review. Your goal is to persuade this company to adopt better procedures for handling cash. After assessing your audience, you feel that as a group they are unfamiliar with the procedures you are going to suggest; therefore, in order to persuade them to accept your suggestions, you will have to spend some time during the presentation explaining the basics of the method as well as providing reasons why this method is the best choice given their circumstances. You should plan to fill the gap between what is known and what is needed by providing optimum, not maximum, information. For example, explain only relevant procedures for this situation, not all potential controls that could be considered in a broader context.

WHAT ARE MY AUDIENCE'S FEELINGS ABOUT THIS SUBJECT?

Suppose you need to write a letter reprimanding a staff member, perhaps by critiquing unfavorable segment performance against its budget. Such communication situations are sensitive because your message could trigger emotional responses in your audience. While writing a reprimand, remember that few people enjoy being criticized and often react defensively. If your goal is to provide constructive criticism with the hope that the individual will learn from mistakes, you first have to break through the resistance that usually develops when someone feels attacked. Taking into account how your reader is going to feel about this letter will help you to create a text that acknowledges those feelings but focuses on the opportunity for improvement. Presenting an evaluation of performance against budget involves many of the same issues. You want to point out problems and weaknesses you noted, but you don't want to alienate your audience. If the objective is to promote corrective and adaptive action, do not come across as if you are there to place blame.

Since emotions often filter or deflect the intended message, writers and speakers should attempt to take the reader's or listener's perspective. Imagine opening and reading that letter of reprimand. Imagine receiving a critique of your segment's performance—not as *you* personally would read it, but instead as the critiqued persons, with *their* feelings, goals, and jobs in mind. Their perspectives include the overtime, the sweat, and the pride they invested in their work, all factors perhaps invisible to you. When you try to consider communication from your audience's point of view, you can better anticipate emotional minefields and plan your message accordingly. Remember that in this imaginative identification, which some therapists and social scientists have called *empathy,* your own biases are still at work. Thus, just because *you* would rather be complimented constantly throughout the day, do not assume that others have the same preference. Although it sounds nice, "Do unto others as you would have them do unto you" may at times be poor advice for communicators if the "others" do not like the same kinds of "doing unto."

The examples so far highlight situations where it seems obvious that feelings would be affected. In many situations where emotions are less obvious, they are just as important. For example, if you are analyzing which of several mutually

PRACTICE

Considering the Audience's Feelings

The Athics Company purchased a machine a year ago for $500,000. Originally, the company decided to depreciate the asset over eight years using the straight-line method. The chief financial officer of the company has been discussing changing the depreciation to a 12-year life. He is concerned about current earnings and sees this as a way to lower expenses with no harm to the company. You are concerned about this change since you believe that the useful life of the asset is unaffected and that this proposed change would be unethical.

Task: Write brief memoranda using the following pairs of roles and audiences. What are the differences in what you said and how you said it, given your role and the audience?

Role (You)	**Audience**
Assistant to chief financial officer	Chief financial officer
Staff member of firm that audits Athics	Your superior
Staff member of firm that audits Athics	Chief financial officer
Partner of firm that audits Athics	Chief financial officer
Vice president of operations: Athics	Chief financial officer

exclusive projects to recommend to operating management, remember that each project might have its own champion in upper management or might have been a pet project of a lower-level manager. You cannot ignore the interaction of what you say and how you say it in relation to feelings.

WHAT CULTURAL DIFFERENCES MIGHT INFLUENCE HOW MY AUDIENCE WILL READ OR HEAR MESSAGES?

In this era of growing international trade, you could easily find yourself dealing with executives from other countries who have alternative ways of communicating within their cultures. You will be communicating using the same language (perhaps English), but you need to be aware of possible problems that can occur with nonnative speakers of English. Many phrases Americans use are often misunderstood by people for whom English is not their first language. These phrases, called *idioms,* do not mean exactly what their words imply. Examples of idioms in American English include "kick the bucket," "deep six the report," and "a Catch-22 situation." If you know your audience will include nonnative English speakers, you will need to plan your communication carefully and either avoid or briefly define words and phrases that could confuse listeners or readers.

Nonverbal communication also has its own set of problems. In the United States, touching your index finger and your thumb together to form an "O" is a sign for OK. In other cultures, it may signal a person being a *zero* or even worse.

Cultural differences are not limited to international communication. Many cultures, including ethnic and religious groups, coexist within the United States. Most people are sensitive to their own cultural group but may forget their group contributes to a multicultural society. As a professional, you need to be aware of and sensitive to cultural diversity. For example, it is common in some Eastern and Asian-American cultures for people to act as if they are agreeing with something even when they are unsure or really disagree; it's a matter of manners. In all situations, by realizing how your audience may be affected by your words, actions, and style, you can make better decisions as a communicator.

Audience Analysis Strategies for Written Communication

Writers have a major disadvantage in adapting to their audiences. Most writing is not read in the presence of the author. This naturally results in a lack of immediacy; as a writer, you cannot assess your readers' reactions as they read. You will not be there to answer questions, nor will body language and voice inflection clarify your message. Readers can work only with the text in front of them and cannot read your mind. The written word often gets only one chance to be "clear"; your thoughts are represented only by the words on the page. No other factors can come into play. You can edit and rewrite something until you are satisfied with it, but once a written message is delivered, you have relinquished control over it.

This lack of immediate feedback challenges writers to create accessible texts that lead readers through ideas clearly and logically. Once you have analyzed your reader's needs through the use of the six questions discussed earlier, you also need to be aware of the importance of the surface conventions that are vital for communicating effectively. These surface conventions include paragraphing, spelling, and punctuation. Even the most careful audience analysis will be of little use if you create ambiguous sentences or if you fail to order your information into logical paragraphs. The reader needs to know where one sentence stops and another begins, where one idea stops and a new one begins, and how your thoughts are related to one another. A more complete discussion of surface conventions is included in Chapter 6.

Here then are strategies you can use for written communication, which reinforce the general principles of audience analysis.

1. Determine as nearly as possible who your target audience is.

2. Use the audience analysis questions listed earlier in this chapter.

3. Form a mental picture of your audience as you begin to write. Throughout the writing process imagine that you are conversing with your readers even if that conversation sounds fairly formal at times.

4. Decide on the best possible written format for your audience. Will your readers respond best to a formal letter, memorandum, e-mail, or report? Can your message be given informally? Does your audience need extensive supporting materials (graphs, spreadsheets, etc.)?

5. Determine what tone you wish to establish in your text. Do you want to be friendly and folksy? Do you need to remain more distanced from your readers through the use of more formal language? Tone is established through word choice, and once you have decided on the tone you want your writing to have, you need to maintain that tone throughout the text.

6. Decide on the style of writing that best suits your audience. Again, you can choose between informal and formal styles. An informal style can include using contractions and a more conversational approach. A formal style would be less personal and follow the conventions of business writing. Perhaps most accounting audiences would respond best to an economical technical writing style.

AUDIENCE ANALYSIS STRATEGIES
FOR WRITTEN COMMUNICATION

- Determine your target audience.
- Answer fundamental audience analysis questions.
- Form a mental picture of the audience.
- Determine the best possible format.
- Determine the appropriate tone.
- Determine the appropriate style.

Audience Analysis Strategies for Oral Communication

Remember that *immediacy* is the basic distinction between written communication and face-to-face speech. Although most basic audience analysis strategies apply equally well whether writing or speaking to audiences, conversational or public speakers must account for this added subtlety. Unfortunately, most speakers ignore the potential for audience analysis in the immediate experience of speaking.

Think about which audience analysis factors are most relevant to the spontaneity of a particular public speaking situation or an interpersonal conversation. Consider what you can do to maximize your effectiveness on the spot. The following are some useful strategies.

1. Recall what you know (if anything) about audience members' previous habits in other speaking situations. You may be able to infer what kinds of reactions are more likely when you speak.

2. Develop contingency plans based on different possible reactions from the audience. Do not rely on the audience or any of its members to react in the particular way you expect, and for which you have planned. Your funniest story may be met with blanketing silence; your question-and-answer period may stimulate zero questions; your impassioned call for a volunteer may be futile. A more positive aspect of this suggestion is that speakers who know ahead of time that they will be speaking to an enthusiastic and responsive group need to anticipate how they will control the time or put the discussion on track if the audience enjoys the ideas too much and begins to run with them.

3. Monitor the listeners' nonverbal responses while you speak to obtain cues for adjusting the remaining parts of your message. This advice is easier to give than to practice, and even a detailed summary of the "hows" of such nonverbal sensitivity is beyond this book. Entire books are devoted to analyzing nonverbal behavior as cues to human emotion; however, below are some useful, if overgeneralized, indicators.

   —Eye contact in general North American culture usually signals the attention (though not always the interest) of the listener. Eye contact is less reliable as an indicator of attention among some cultural groups (for example, among many Asian-Americans and Native Americans) because for many people averting the eyes is a sign of respect.

   —Posture and movement can be reasonable indicators of audience attention. Generally, interested listeners tend to exhibit a forward-leaning posture facing the speaker and do not fidget.

   —Facial expressions, though often ambiguous, can be helpful to monitor. Instead of simply noting whether people are smiling or frowning as positive or negative feedback, be aware of when they smile or frown. Bored listeners can paint smiles on their faces as façades, and fascinated listeners can frown as they struggle internally with a challenging idea that

ADDITIONAL AUDIENCE ANALYSIS STRATEGIES
FOR ON-THE-SPOT ORAL COMMUNICATION

- What do you know about this audience from previous experiences?
- What will you do given different audience reactions (contingency plans)?
- What cues are you picking up from the audience as you present? eye contact? posture? noise or silence?

fascinates them. What you want to check is whether the smiles correspond to the humorous sections of the talk, and whether the frowns might be a generalized negative reaction. Remember, too, that few audience members are totally in control of the appearance of their own nonverbal cues and may disclose their emotions in quite varied ways.

—Silence can have positive as well as negative dimensions. Silence in response to your requests for involvement or interaction may mean that your audience couldn't care less, or it may mean that listeners care so much about your message that they want to give it some serious thought. Serious thought takes time. Do not automatically assume your audience is alienated if they are silent.

PRACTICE

Putting It All Together

You are the new chief financial officer of a company that has just been bought from its parent. The board of directors of the new company consists of hourly employees of the company, members of management, and some outside investors. You have been reading some literature saying that it's important for everyone from the board down to hourly employees to know how to read financial statements.

Tasks

1. If you were making a presentation to the board of directors instructing them how to read financial statements,
   a. Would you want to do it in writing or orally? Explain.
   b. What would be an appropriate approach (plan) for such a presentation? What is your reasoning?
   c. What visual aids would you want to employ? Why?

2. If you were making a presentation to hourly workers in the factory instructing them how to read financial statements,
   a. Would you want to do it in writing or orally?
   b. What would be an appropriate approach (plan) for such a presentation?
   c. What visual aids would you want to employ?

You may protest that all this on-the-scene analysis is hard. You could ask questions like: Isn't it likely to take my mind off my own thoughts? Won't I lose my place? Won't I forget where I'm going with my ideas? The answer to these good questions is simple, but hard to swallow for a speaker who is not willing to practice.

Yes, immediate audience analysis will deflect you from developing your own thoughts if you are inadequately prepared or if you are not self-confident in the first place. Speaking with a group, or conversing with another person, for that matter, is very much like improvisational jazz. It does not make sense for a saxophonist to say to a pianist, "I wish you would not play while I'm trying to create my own music; it distracts me." The basis of improvisation is playing off the ideas of others. Without that sensitivity, that flexibility, it's not jazz. To extend the analogy to communication, there are no solo acts involved in speaking to a group. Everything is ensemble work. Without the flexibility of audience awareness, it's not effective communication. The success of the communication is the success of the speaker and listener(s) creating meaning together. As a speaker, you must practice not only how to prepare and express ideas but also how to listen and monitor how those ideas are growing or deteriorating among your "musical" partners.

Summary

Audience awareness is perhaps the most crucial factor in the success or failure of any communication plan. Effective communicators never forget their audiences, and they use their audience's needs as the major component in all the decisions they make as they plan for communication. The people you interact with expect and deserve accessible and comprehensible messages.

How to Approach Practice Tasks

LOOKING AT A WRITER-CENTERED TEXT

What's important here is to see things from your intended audience's point of view. Terms familiar to you (jargon, abbreviations, tone, etc.) might be unfamiliar to others. Content changes, too. You might include some private thoughts and feelings in material you wrote for yourself that are either not appropriate for others to see or would make you vulnerable to others.

REWRITING BASED ON WHAT YOUR AUDIENCE KNOWS

In each case, the audience is not familiar with the technical jargon included in the original message. You will want to construct your response by taking into consideration the frame of reference (knowledge, experiences) of your target audience. Here are some suggestions.

- **Production managers:** In order to comply with accounting rules, we divide costs into two pools: costs associated with everything you and your people do in the plant and costs associated with the front office and sales department. When we look at the cost of any particular product, for example Model 43093, we look at the cost of materials and components (like the housing) we buy and the costs of the hourly labor for making the product. In addition, as you all know, there are a lot of costs out there that are not just in materials and labor. For example, John's department maintains all the equipment and Sally's does all our in-plant engineering. We look at how those costs are associated with each product and give each product a share not only of these specific costs but also of general costs like heat, light, building insurance, and so on. All of these supporting costs are called *overhead.* Thus, when we build 100 units of Model 43093, we take the costs associated with these 100 units (materials/components, labor, factory overhead) and associate those costs with those units. Let's say the total is $30 per unit. If 100 units are in inventory, then they are valued at $3,000. Later, if 50 are sold, inventory is lowered to $1,500, while $1,500 appears in the income statement as the cost associated with sales of 50 units. Let's compare this to sales costs; these costs are part of the monthly cost of doing business and appear in the monthly income statement. These costs are not related to whether a product is kept in inventory or sold.

- **Sales managers:** In order for you to know more about how we set prices, let's talk about how we divide our costs for financial accounting. All costs are associated with either what we do in the plant or what we do in the front office or in sales. Your salaries and other costs are expenses each month; we lower income given these expenses. In contrast, costs associated with making our products only affect income when a unit is sold. As you know from talking to the estimators in the plant, basically our product costs are the materials and components we buy, the labor we use to make and assemble the products, and the general plant costs, called *overhead,* that we assign to each product.

- **Union officers:** I know you are all interested in how we look at costs here in the plant. Part of how we do things is related to what we have to do to create financial statements for the banks, our owners, and the IRS. Rules call for us to divide our costs into those in the plant and those in the front office and sales. Let's look at plant costs. For example, how about Model 43093, which is made in part in several departments. We take all the material and components (like the housing) we buy and add on the costs of the union hourly workers. You know that your people record the hours they work on each job. This is part of how we can see what a job costs. We also add the costs associated with almost everything else in the plant. Thus, we take some of Joe's time in quality control, Pete's in managing the racks where we store materials, and Sarah's for moving stuff with the forklift and add that to the cost for every product. In addition, some general costs like heat, light, supervision, and so on, are added. All these supporting costs are called *overhead.* When we're all done, we have the cost of each product consisting of its materials/components, labor, and share of overhead. When we make 100 units of Model 43093, we can then value them at that cost whether we keep them in inventory or sell them.

Here is where the big difference comes in with the plant costs and the other costs. For our income statement, we show only plant costs associated with the products we sell. However, for the front office and sales costs, everything we spend each month is shown as an expense for that month.

- **High school students:** As accountants, we are governed by rules on how to deal with information. One of these rules requires that we divide costs into two parts: costs associated with making our products and other costs (like the general management of the company and sales). You are all familiar with textbooks, so let's use them as an example. If we were a company printing and binding texts, we would keep track of the costs of the paper, ink, cover material, glue (all generally called *materials*), the labor associated with the people on the presses and binding machines, and all of the other plant costs. Taken together, we can come up with a cost per book. The difficult part is assigning general costs (like the cost of supervisors, engineering, heat, light, and plant insurance—collectively called *overhead*) to each book. Some of these items are easier to associate with the book (like maintenance of the machines on which it is printed), while some are almost impossible to trace (like light or the plant manager's salary). However, we find a way to assign all these costs. Thus, whatever books are in the plant are valued at a cost consisting of materials, labor, and overhead. When the company sells books, those costs are shown as an expense. Thus, the company would show revenue for the sales less the costs of the books sold. In addition, the company would also deduct costs associated with the general running of the business (administrative and sales costs).

As you read the above, you can see that parts of each might be used for the others. These are just samples of what you could do. How do your statements compare? Where do you believe yours are better? Not as good? Why?

CONSIDERING THE AUDIENCE'S FEELINGS

Remember that your role and your relationship to your intended audience will affect how you develop your message as well as how you are likely to be perceived. Since a question of ethics is involved here, you will also need to be sensitive to shaping your message to avoid being accusatory or confrontational. As assistant to the CFO, several issues must be considered. How will this decision reflect on the department? If an outside auditor comes in and tells upper management that this is improper and, more strongly, a manipulation of income, how will this reflect on you as someone who has probably created support material at the direction of the CFO? Thus, there are feelings and potential legal issues to consider.

When you move to the role of independent auditor writing to your supervisor, you can be much more blunt. State what you think is going on here. Also remember that what you write can be part of a public record if any legal action is taken.

The bottom line, then, is that this example involves more than feelings (personal investment in the decision) and could involve legal issues.

PUTTING IT ALL TOGETHER

The format of your presentation for the board of directors will be determined by the expertise of the people who constitute this group. If, as a group, they lack experience with financial statements, you will probably want to use an oral presentation. This format will allow for questions and discussion. A more sophisticated group will probably respond well to a simple written presentation. In both cases you will want to include visual representations of financial statements with which you can point out specific sections or details. Graphs are also appropriate.

The hourly workers will most likely respond better to an oral approach, enhanced by handouts or overheads that depict specific details. You can plan to be a little more informal with your approach to this group. Graphs might replace numerical tables. Remember your goal is to provide complex information for a nontechnical audience. Provide the basics without overwhelming them with unnecessary detail.

Chapter 4
Determining Approach

Y ou have been asked to analyze the possible purchase of a large piece of equipment costing $6,000,000. Your report will go to the vice president of operations, and she will then generate a recommendation to the president. There are several ways to approach this task. You could start with the marketing people to get information about customer needs and projected sales to utilize the equipment. You could go to the Internet and see what you can find out from competitors', manufacturers', and customers' home pages. Another approach would be to start with operating plant personnel and the engineers and begin estimating operating characteristics and costs. There are other ways to start—each with its own advantages and disadvantages, each possibly leading down different paths toward your final analysis.

In fact, your analysis might lead to a partial redefining of the task. While you are interested in whether to buy the equipment, you might find some interesting aspects of the competitive environment that you want to consider, even though they go beyond your original purpose. Reviewing *The Wall Street Journal Interactive* Web page could reveal the recent merger acquisition of a company that will become a segment of a competitor and produce products in a closely related field.

This example illustrates an important aspect of communication planning: approach, the final ingredients of any communication plan. You will consider the types of approaches you can use for conveying your information, possible organizational schemes and formats, the process of incorporating information from other sources into your own message (synthesis), sources for researching your topic, and choice of graphics to support your information. In this chapter we do not separate strategies for written communication from those for oral contexts, because the same questions serve to guide your content decisions as both a writer and a speaker.

Focusing on What You Have to Say

In previous chapters we looked at two vital parts of a communication plan: purpose and audience. Knowing your intended outcome and being aware of your intended audience are essential before you can begin to create and shape your message. However, before you begin to draft a letter, report, or presentation, you need to make some decisions about the content of your message and how best to focus what you have to say. During this stage of planning, you first need to determine what you might want to include in your message.

- Do you have the information you need, or should you do some additional research?
- What is the best organizational scheme to use?
- Is your information best represented visually, and if so, what sort of graphic presentation will work best?

All of these aspects need to be considered as you prepare to put pen to paper or words on screen.

Thinking through this part of your communication plan may take only a few minutes while you jot down notes and reminders to yourself, or it may involve several weeks of gathering and preparing data. Determining content and approach is just as important as understanding purpose and analyzing audience in the overall success of your communication. Thus, you will want to approach every communication situation with a clear sense of purpose, audience, and content (a *communication plan*) before beginning the process of drafting.

Deciding on an Approach

Choosing an approach is much like a manufacturer choosing a package for a particular product. In many cases, how the product is packaged will influence how the public perceives the product and may affect its overall acceptance. The same is true for communication. The way you decide to *package* your information will greatly influence the outcome of your message. Although packaging a message includes details such as paper quality or letterhead design, we are going to concentrate on the kinds of rhetorical approaches you can take in presenting your

---

THINKING ABOUT WHAT YOU WANT TO SAY

- Knowing your intended outcome and being aware of your intended audience are essential before you can begin to create and shape your message.
- Determining content and approach are just as important as understanding purpose and analyzing audience in the overall success of your communication.

message. These approaches are not mutually exclusive; rather, you will make choices that combine elements of all of them.

DISTINGUISHING ANALYSIS FROM DESCRIPTION

Many people have a difficult time distinguishing between description and analysis as approaches to communication. Although some communication contexts may require that you provide descriptive information, most of the communication situations accountants (both students and professionals) encounter involve analyzing information.

The difference between description and analysis is the difference between recapping and probing. It's the difference between writing a pure statement of facts and writing an investigative report. When you are using a descriptive approach to communication, you are simply giving a recap of the data; perhaps you are summarizing last month's financial results. However, analysis necessitates that you probe into the information enough to discuss its meaning. This probing is accomplished by answering the questions Why? How? What are the ramifications of this information? and So what? When you provide answers to these types of probing questions, you have entered the realm of analysis. A mere description may not satisfy your readers or listeners. By shifting to address the reasons why last month's results did or did not meet expectations, you are providing an analysis of the situation.

For example, suppose a company's accounting system yields the following information:

	Budget	**Actual**
Sales	$1,000,000	$ 900,600
Units	100,000	94,800
Price	$ 10/unit	$9.50/unit

A *description* would be:

> The budget called for 100,000 units to be sold at $10 each or total revenues of $1,000,000. Actual sales were $900,600 with 94,800 units sold at an average of $9.50 per unit.

Analysis could include:

> Sales fell short of expectations both in price and in quantity ($9.50 versus $10.00 per unit; 94,800 versus 100,000 units). This was caused by a price war in the industry. Our volume went down before we could react in the marketplace by lowering our own prices. Thus, both volume and average price suffered. When we are faced with this kind of competitive pressure in the future, we will have to take more immediate action. On the plus side, costs were in line with expectations. Thus, income was affected only by the shortfall in volume and average price.

Many complex texts (reports, proposals, or case studies) require both description and analysis. Without both, the audience is left hanging and will no doubt come to its own conclusions about the information you have presented as well as its own conclusions about the quality of your thought and effort.

Analysis versus Description

United Airline's home page (http://www.ual.com/asp) shows the following information about some of the airplanes previously used by United.

Airplane	First Year in Service	Passenger Capacity	Cruising Speed
DC8	1959	114–149	545–600 mph
B720-022	1960	115	600 mph
SUD Caravelle	1961	64	500–530 mph
Vicker's Viscount	1961	44	350 mph
B727-22	1963	90	550–600 mph
B737-222	1968	91	500–550 mph
B747-100	1970	370	600–625 mph
DC-10	1971	222	600 mph

Tasks:

1. Write three facts illustrated by these data (description).
2. Name three groups that might have an interest in these data.
3. Write a brief analysis of the data for each of the three groups you listed in Item 2.

Let's talk about cases more since they are simulations of situations you will encounter as a practicing professional. Recall from our discussion of role, audience, and task that some cases are explicit about these factors while others are not. However, in most instances, the objective is analysis. Depending on role, audience, and task, an instructor may or may not want description (e.g., summary of relevant facts). Usually, you are asked to focus on analysis, problem solving, and decision making. These ideas are expanded in Chapter 7.

FORMAL VERSUS INFORMAL APPROACH

Not all the writing or speaking you will do needs to be as formal as an article in *The Accounting Review* or *Harvard Business Review*. Deciding the level of formality of your approach involves a choice between formal or informal language that will determine the overall tone of the text.

Tone can be suggested in both written and spoken language. In written language, tone refers to the overall feeling evoked by the words. In oral language, tone is further demonstrated by the ways the words are spoken; pitch, rate, volume, and inflection all contribute to perceived formality. A formal tone is created by using words and sentences that create a distance from the reader. Formal language avoids contractions and usually does not address the reader directly. It may rely more on passive voice ("It was determined that . . . ") than active ("The comptroller discovered . . . "), although formal language can also include the active voice. This is the language of an accounting standard or of audit reports. *Informal language* is more like the language of everyday conversation. Although it will not usually include

Formal versus Informal Approach

In which of the following role-audience-task combinations would you choose to use a formal approach and in which an informal approach? Clearly state why you have made your choices.

- You are the speaker at the local meeting of the Association of Government Accountants. You are giving a presentation about new software available in the field.
- You are the local plant manager of a multinational firm. The board of directors is having its meeting at your plant this month. You have been asked to make a presentation regarding your plant's operations.
- You have been asked to be a professional witness in a utility regulation case. You are to write some testimony dealing with the relevant issues of the case.

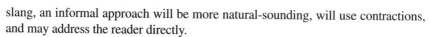

slang, an informal approach will be more natural-sounding, will use contractions, and may address the reader directly.

Remember your audience analysis when making formality decisions. As a communicator, you will need to decide whether your message is best related to your particular audience through formal or informal language. Your decision about formality will also be influenced by the communication context. Some organizational cultures and situations demand formality, while others do not.

DIRECT VERSUS INDIRECT APPROACH

A third decision about approach involves the degree of directness you wish to employ. When you are analyzing sensitive areas in which people have strong feelings, how direct you choose to be may be crucial. For example, reports that include criticism, suggest any sense of bad news, or involve pet projects, pet peeves, or favorite topics of some readers or listeners are particularly touchy.

In such situations, you have the choice of being direct with your communication or approaching the message in a more roundabout (indirect) way. There are sound reasons for using either approach. At times your decision will be influenced by the expectations of your audience; other times you will choose between direct and indirect based on your comfort with one of these approaches. However, even if you are more comfortable with one approach or the other, you must be aware when each is appropriate. The difference between the two approaches is apparent in the synonyms used for each.

Direct is also blunt, candid, specific, frank, and plain; indirect is circuitous, oblique, roundabout, and vague. When you are direct, the audience knows as soon as possible exactly what you are thinking and why you are thinking. In an indirect approach, the audience has to infer more from what you are saying. Consider the following examples of direct and indirect approaches:

- **Direct:** The manager of this project seems more interested in preserving his image as a successful leader than in looking at the reality that conditions have changed with the project and it should be abandoned.

Direct versus Indirect Approach

You have been part of a team assessing the effectiveness and efficiency of the vice president of finance's office. His duties include not only generation of external reports to shareholders and taxing authorities but also the coordination of information for management decision making. Your team has concluded that the vice president has not been doing a good job in providing managerial information, while he has been doing an adequate job in his other responsibilities. In addition, you have sufficient evidence to conclude that he wants to have as much personal power within the organization as possible. He wants to be seen as indispensable and the confidant of the president.

Task: Using the guidelines and the illustrations for direct and indirect methods, write two to three sentences using each of these two approaches. Do not use the same content or wording as the example in the text.

- **Indirect:** Good decision making and ultimate success involve knowing when to cut bait with a project that does not seem to be making it.

The first example illustrates a direct criticism of the project manager. This might be appropriate in a case analysis where you are a student writing to a teacher; there are no egos to hurt in that situation. If you were writing a report evaluating performance given certain criteria, you might choose the direct approach and then support your evaluation with reasoning and evidence. For a report dealing with general operations, especially one that will be read by many people, the indirect approach would seem better. Note that our illustration of the direct approach is strongly stated. You can be as direct in many other ways.

As you can see from this example, the indirect approach often leaves out particular information. In this case, direct criticism of the manager might not be useful. One side effect of saying things indirectly, though, is leaving your readers or listeners to wonder if there is some hidden meaning in what you are saying. Thus, they might reach a wrong conclusion about your conclusions.

Many of us choose to use a direct or indirect approach based on the way we like to receive sensitive news. This can have its good and bad points. While it's important to think about how you would feel if you received information using the direct or indirect approach, it is also important not to let your own biases influence what is appropriate in a situation and how you should think about the audience's needs and feelings.

Chapter 7 discusses some specific formats (e.g., reports, letters, memoranda) in more detail. However, examine a couple of these in light of this discussion. If you were writing a letter to a client regarding the application of an accounting rule, your approach likely would be fairly formal and direct, including both a description of the problem and your analysis of suggested solutions. Compare this to an internal memorandum dealing with a sensitive subject. If this memo is

---

WHAT DOES THE TASK CALL FOR?

G iven your role, audience, and task, choosing an approach for either oral or written communication means you have to determine whether your task primarily calls for

- Description or analysis.
- Formal or informal tone.
- Direct or indirect expression.

---

addressed to your subordinate, you might be fairly informal, describe the problem at hand, and choose to be indirect depending on the sensitivity of the issue.

These same concerns and choices apply to oral communication as well. A report to the board of directors on next year's plans might be formal, direct, and both descriptive and analytical. If you were giving instructions regarding the content of a training seminar, you might be informal, direct, and mostly descriptive. As you can see, the combination of choices you make really is specific to the situation at hand, whether you are dealing with an oral or written context.

Choosing an Organizational Scheme

Once you have thought through your combination of choices in terms of approach, you are ready to think about organizing your information for a clear presentation.

THE THREE S'S

A primary organizational scheme is one that can be used in almost every communication context from short memos to long speeches. It involves dividing your presentation or text into three steps or segments:

Say It
Support It
Say It Again

When you use this pattern, you use the opening section (introduction) to present your most important point(s), you use your middle sections (body) to support or illustrate your ideas, and you finish your communication (conclusion) by restating your main point(s). Practically speaking, this format introduces important ideas up front and closes by reaffirming those same main points. Of course, the "Say it again" advice does not mean to present the ideas verbatim. A conclusion that basically just repeats the introduction will insult readers and listeners. Think of introductions and conclusions together as the frame around the picture of your text. They should complement one another as well as structure the overall message clearly for your audience. The introduction provides a context for the communication and is where you state its purpose (thesis). In the conclusion, you show how you achieved your purpose and provide a context for whatever might follow (e.g., additional analysis, application of a particular accounting rule). While the

three S's are important in both written and oral communication, they are especially useful in longer oral settings since listeners, unlike readers, are unable to go
back to a previous page the way readers can.

STATEMENT OF PURPOSE AND CONTEXT

No matter what the organization or content of the material, considering your audience means you should ensure that they know what your communication is all
about. In the three S's, part of the first "Say it" is establishing purpose and context. You are asking people to use part of their busy schedule to read or listen to
what you have to say. They should know up front what value you are adding.

Purpose. Know your purpose. It might be a description of an existing situation (just putting some facts on the table). It might be an analysis. If you reach
conclusions or recommendations, let your audience know what they are.
Establishing purpose is a different task than providing an executive summary
(discussed in Chapter 7). Think of purpose as the reason why your audience
should pay attention to what you have to say and the foundation upon which you
build your organization and content. Without a clear sense of purpose, communication will not be effective.

Context. Any audience has to be brought up to speed. This might take a sentence or two or longer depending on the audience and the task. A report (oral or
written) should be put into a context so that the audience knows how it fits. Your
role might be to synthesize several background reports and to provide recommendations. Part of providing context for your audience would be to remind them of
what has preceded your report. In addition, if there are logical steps that will follow, let them know the future context as well.

   If someone reviews a draft of a written or oral report and finds the overall
organization unclear, this is often caused by ill-defined context and purpose.
Before you can really address your task and achieve effective communication,
you have to think about, draft, and redraft your statement of purpose and context.
As you develop the content of your presentation, keep coming back to that statement of purpose. Keep asking yourself: Is what I am saying here helping me to
achieve my purpose? How is this linked to my purpose?

ORGANIZATION OF THE BODY OF THE MESSAGE

Topical organization. Organizing ideas by subjects or topics is common and
clear to most readers or listeners who are familiar with the terminology you use to
identify the topics. In preparing a report dealing with your firm's new corporate
headquarters in Detroit, you may wish to present the implications of the move in
four parts: (1) changeover costs, (2) cost of living differences, (3) availability of
materials, and (4) corporate image. If discussing the adoption of a new Financial
Accounting Standards Board (FASB) rule, topics could include (1) related rules,
(2) points of view raised by public accounting firms and/or major corporations,
and (3) projected impact, for example. Take care not to divide your organization
into too many subsections; three or four divisions should be appropriate for most
topically organized brief reports or presentations.

Chronological order. When you order your information chronologically, you are presenting the information in relation to time. You take your reader forward through time, detailing events from earlier to later. This organizational pattern works best when you are tracing the history of a problem or are recounting the steps in a process. This pattern is one of the easiest for a communicator to use because the events or steps have already ordered themselves. The writer or speaker does not have to make many decisions about where to start or how to place information sequentially. Using a chronological order is appropriate if you are describing how you have decided to purchase a new division or when dealing with tax cases that are precedents for an issue you are researching. If addressing the adoption of an FASB rule, the time frame of the process is the focus and a chronological order is how you want to organize your discussion and analysis. Be especially careful about using this technique. Because it is the easiest to employ, it is often used inappropriately. Think about the audience. If they do not care about the chronology, or if it is not relevant to their decisions, this may not be a good approach.

Cause-effect or effect-cause. In this ordering system you can begin with a cause and explain the effect, or begin with a discussion of an effect followed by an explanation of the cause. Either way, you are helping your readers to see the relationship between the two. Assume you are dealing with interest rates for loans and how they are influenced by the general rate of inflation in the economy. With *cause-effect* you could say:

> Because the general rate of inflation has remained fairly low over the last few years and is not expected to rise significantly, most lenders are not building an inflation adjustment into the rate of interest they are going to charge.

On the other hand, the *effect-cause* relationship could be demonstrated by:

> Interest rates are going to be fairly constant over the next few years. This is especially true because inflation has remained fairly low and is not expected to rise significantly.

When explaining how the FASB adopts a new accounting rule, you can assume some question has arisen (cause) that led to the discussion and adoption of the new rule (effect).

Decreasing order of importance. This pattern for organization will work when you have several different points to make, but some of them are more important than others. In this pattern you will lead with your most important point and then follow with the rest of your less important ideas. As a communicator, you may want to choose this pattern if there is any chance your reader is not going to read your entire text or if you are concerned that you might not keep your audience's full attention over a long oral presentation. By beginning with your most important point, you can be fairly sure that your audience will remember it. This is called the principle of *primacy.*

Increasing order of importance. Another approach is to use your less important details to build a foundation for the most important point, which you

---

ORGANIZATIONAL SCHEME

1. Say it: Develop context and purpose.
2. Support it: Organize the body of your communication.
  —Topical organization.
  —Chronological order.
  —Cause-effect or effect-cause.
  —Decreasing order of importance.
  —Increasing order of importance.
3. Say it again: Show how your purpose was achieved and provide context for what follows.

---

will deliver last. This pattern often is the choice for persuasive contexts. Assume that you want to convince your audience that the alternative you chose is the best one. You can begin by listing the five options that you considered. Then show why you rejected four of them in the order of the weakest to the strongest of that group. Finally, you show why you adopted your recommendation and how it answers the criticisms you had of the other four. The psychology of this approach, called the principle of *recency,* is that your audience will usually remember the last thing read or the last thing heard. You can see why building a strong foundation and ending with a strong argument is often a good idea. This brings us to two additional topics to include in your organizational scheme: summary and synthesis.

Summary and Synthesis

Beyond deciding on an approach and choosing an organizational scheme, consider smaller refinements like summarizing your analysis and synthesizing outside information. These concepts are useful in introductory and concluding sections of a text, but are important in their own right as part of an overall content scheme or when you are asked to summarize separately from your full communication task.

PREPARING A SUMMARY: ANOTHER PART OF CONTEXT

In some situations you will need to summarize information for your audience. In long reports, this might take the form of an overview called an *executive summary,* described in Chapter 7. However, in shorter reports, a brief summary at the beginning and/or at the end of a presentation can focus or refocus attention on important matters. This is part of providing context to your audience as you begin and as you bring things to a conclusion.

The easiest way to understand a summary is to think of its uses in other more familiar contexts. A synopsis of a piece of literature (such as *Cliff's Notes*) or a sports highlight film provides an overview of the subject being presented. A book synopsis provides only those details that someone who has not read the book will need to know in order to have a sense of the entire text. Likewise, a sports highlight film does not include every play in a game, but only the important scoring plays or turning points in the competition. After reading the synopsis of a book or watching a sports highlight film, the audience will generally understand the book's plot or the flow of the game. By providing a summary, you supply enough information so that others will be able to understand the situation without actually having participated directly.

Once you understand that a summary is an overview of the entire context, you can create a summary by noting all the vital details and giving a capsulized version of their content. Remember that any summary does more than merely introduce a topic; a summary serves as a complete rundown of all the pertinent information.

Your opening should contain a statement of purpose and provide context to your audience. This can include a summary. It can also include a road map of the organization of your communication (another form of summary). Clearly, the conclusion section is a summary of what precedes it; you tie things together to show how your introduction (purpose) was achieved. The body of your paper might also contain different forms of summaries. These might be sections where you condense information rather than provide details or provide a transition to the next section by summarizing what you have said so far.

SYNTHESIS

Many communication tasks require that you work with outside information— that is, data that you glean from sources other than your own experience.

Summarizing

P ick up a recent copy of *The Wall Street Journal*. Read the three main articles that start on the front page. Choose one of these articles.

Tasks:

1. You are the newsletter editor of a publication that summarizes important articles from the financial press. Write a summary of the article you chose to be included in the newsletter. The summary is to convey the full scope of the article. Limit your summary to 150 words. You will also need a headline not to exceed one line. Give your summary to another person who has not read the article. Ask that person to tell you the main point of the article after reading your 150-word summary.

2. You are the news anchor on the local public radio station. Write a summary of the same article you chose to be presented on the radio station. Limit your summary to two minutes speaking time. You will need an appropriate lead-in line. Read your summary aloud into a tape recorder as if you were the news anchor. Then give the tape to another person who has not read the article. Ask that person to tell you the main point of the article.

PRACTICE

Synthesis is the process by which you incorporate this information into your texts. The trick is to blend the resource material seamlessly into your own words while still giving credit to the original source by providing citations. Some communication situations will require a formal citation system, such as using footnotes or endnotes and listing all references on a "Works Cited" or "References" page. In more informal texts, merely mention the name of the source in your text. Most news stories cite general sources or interviews with specific people. Many reports can also refer to general sources or specific people without having a formal citation. For example, if you are referring to the business plan the company has approved, a formal citation is less appropriate than just a reference to the plan using its proper title (e.g., "Our sales goals for this product group are based on the projections included in the FY 2000 Business Plan"). References to people are similar: "Our assumptions about labor skills and costs are based, in part, on our recent conversations with Alvarez and Washington in engineering." Communication involves providing appropriate argument and evidence to support your points (the "Support it" advice we gave earlier). Synthesis from internal and external sources bolsters analysis and conclusions. You are not on your own; you have utilized others' analysis and opinions (affirming and contrasting) to support yours.

When you use information from outside sources, you have the choice of *quoting* the source directly or *paraphrasing* the text of the original. For example, assume you are using an article from an accounting journal, written by Terry Smith. One approach is to quote directly: "Smith states the following: 'It is important for managers to investigate fully the legal, social, and regulatory environment their companies face.'" This can be compared to the following paraphrase: "Smith admonishes managers to look at the full range of legal, social, and regulatory issues affecting business decisions." In both cases, you would have a footnote or other formal citation for the exact source of the quote or paraphrase.

Whether you are paraphrasing or quoting directly, you must give credit to the original source. Beginning writers, particularly, struggle with this requirement. Many students think that if they change a word or two from a reference source, they have put the information in their own words and, therefore, don't have to provide a citation. Remember, even if you modify the wording of a source text, if you are using the unique ideas of another person, you must give credit to that person for those ideas. When you take the words or ideas of another person without giving him or her credit, you have committed *plagiarism*. Proven plagiarism is grounds for dismissal from most universities and certainly has no place in the professional business world either. Thus, the first rule of synthesis is to give credit to your sources.

When you cite sources effectively, you not only give credit to authors but also provide readers with the information they need if they want to consider your sources in more depth. Careful and accurate citations give your readers better access to the information you have used and provide a real service for your audience.

The goal of synthesis is to provide details and supporting material through the use of outside information. When writing or speaking, you want your words and sentences to flow smoothly from beginning to end. You want your information

Synthesis

F AS 128 is a recent change in how earnings per share are reported. At the time of its adoption, there were articles regarding how it might affect investors' perceptions of earnings.

Task: Research two opposing articles from sources like *The Wall Street Journal* or the *Journal of Accountancy,* and write a synthesis (with appropriate formal footnote citations) of the arguments presented.

from other sources to seem to be an integral part of the text and not to stand out as an obvious add-on. Notice how the authors of the following text seamlessly interweave information from outside sources into their writing.

> Allowable pension costs for defense contracts are described in a general way by the Federal Acquisition Regulation System as defined in Chapter II, Cost of Federal Regulations, and in a more specific way by two CASB standards relating to pension costs: CAS Nos. 412 and 413. The CASB standards define allowable pension costs in ways that are generally consistent with tax rules, but are slightly more stringent than ERISA rules. The CASB guidelines are also very similar to Generally Accepted Accounting Principles under APB Opinion No. 8 [AICPA 1966], with one important exception: only amounts *contributed* to the fund by the date established for filing a federal income tax return are allowed.[1]

Although the text includes acronyms likely to be understood only by professionals in the field, it is clear in relating facts and ideas to each other and in spotlighting sources a perceptive reader may want to investigate further.

Sources of Information

Effective content relies on determining relevant sources of information. Many other sources are available to supplement organizational records and the knowledge you have acquired in your position. Internet and database technology has enriched the whole notion of sources in recent years. In some ways, we are faced with a haystack of data and must find the proverbial needle of relevant information within. Because it is relatively easy to get on the Internet and rely on powerful search engines, it is tempting to ignore more traditional sources, such as library resources and actual hard copies of information (journals). Be careful that you do not use Internet sources as a mere shortcut.

HARD COPY INFORMATION

There is a fine line between what we think of as hard copy and what is available via computer (CD-ROM, DVD, Internet). Much of what is available in printed form can also be found in electronic form and printed by you as needed. We offer some specific suggestions about Internet sources in the next section.

Company information. Most public companies will send on request copies of recent annual reports. You might also be able to get other information such as their mission statement and basic goals and objectives, copies of company publications for their employees or for the trade, and other documents. Other sources include Moody's Financial Services manuals, Standard & Poor's *Standard Corporation Descriptions,* Dun & Bradstreet's various print and electronic sources of company information, and various Hoover's materials in handbooks and on-line.

Industry data. Most industries have trade associations that publish newsletters and magazines to provide information about the competitive environment, changes in technology, new products, salaries, and industry trends. In addition, your local library may subscribe to the periodically updated *Encyclopedia of Business Information Sources,* published by Gale Research. There is also *Standard & Poor's Industry Surveys* (a quarterly reference service) and the newly revived *U.S. Industry & Trade Outlook.*

Publicly available data. Information on organizations with publicly traded securities is available in hard copy formats such as microfiche, as well as from many computer sources such as LEXIS/NEXIS, Compustat, Prodigy, and CompuServe. In addition, some of these services are offered on CD-ROM. There is the *Accounting & Tax Index,* which is available in hard copy or electronically, and the electronic National Automated Accounting Research System (NAARS). You can find daily stock prices, 10-K's, annual reports, and other publicly available information from these sources in complete or summary form. In addition, with computer sources such as these, you can search long documents for key words. For example, if you are interested in what companies were potentially affected by a major natural disaster, you can search LEXIS/NEXIS for all annual reports that mention the occurrence.

Professional and academic studies. Journal articles and books are good sources of information gathered from studies of specific topics, companies, and industries. Computerized periodical indexes such as ABI/Inform allow you to use key topic words to find potential sources. Researchers in many areas of accounting are publishing more field studies in monographs and in accounting journals, providing interesting sources for in-depth understanding of what is happening in a particular company or industry. In addition, large accounting firms are a good source of information, as they often perform studies of a particular industry.

Professional organizations. The American Institute of Certified Public Accountants (AICPA), Institute of Management Accountants (IMA), Institute of Internal Auditors (IIA), and Financial Executives Institute (FEI), among others, publish journals, newsletters, monographs, standards, and so forth. These organizations have catalogs describing various sources of information.

Governmental data. The U.S. government publishes a myriad of information sources about varied topics. For example, you can find housing starts, automobile sales, beer consumption data, and various indexes (e.g., Consumers Price Index).

Most data collected by governmental agencies are available in microfiche or original text. The Congressional Information Service in Washington, D.C., publishes the *American Statistics Index,* the *Index to International Statistics,* and the *Statistical Reference Index,* all of which collect information available from the state, local, and federal agencies as well as international organizations. You may want to consult the *Monthly Catalog of United States Government Publications* from the U.S. Government Printing Office. It is referred to as *GPO,* and there is a *GPO Access* page on the Internet from the Superintendent of Documents home page.

Generally Accepted Accounting Principles. Besides textbooks, the Financial Accounting Standards Board documents the various positions taken by interested parties when new accounting regulations are being discussed. Large firms might also have position papers they would send to you.

Tax regulations. Tax research includes various books as well as computer sources dealing with the tax code. For most beginning researchers in tax regulation issues, looking at hardbound books is a much more efficient and effective way of finding information. Much tax research involves knowing specific jargon; thus, most people would be limited at least at first in trying to use computer sources where the proper key words are needed for access.

INTERNET SOURCES

What follows is a sample list of Internet sources. Since the list changes daily, this is provided as just a core rather than an exhaustive set of sources.

Site	URL
Accounting firms	Use Rutgers Accounting Web below
Accounting professional organizations	Use Rutgers Accounting Web below
Accounting System Locator/Selector	http://www.excelco.com
American Institute of Certified Public Accountants	http://www.aicpa.org
Annual Reports Online	http://www.zpub.com/sf/arl-www.html
Economic statistics	http://www.bos.business.uab.edu/data
Federal Reserve Bank of St. Louis	http://www.stls.frb.org
Financial Accounting Standards Board	http://www.fasb.org
Government Accounting Standards Board	http://www.gasb.org
Internal Revenue Service	http://www.irs.gov
International Accounting Standards Committee	http://www.iasc.org.uk
Rutgers Accounting Web	http://www.rutgers.edu/Accounting
Security and Exchange Commission	http://www.sec.gov
Stock quotes	http://www.stockfind.newsalert.com
The Wall Street Journal	http://www.wsj.com

In addition, most companies have their own home pages. These include their annual reports (which can usually be downloaded using free software, such as Adobe Acrobat Reader) as well as information on human resources, customer service, and so forth. Search engines such as Northern Lights (http://www.nlsearch.com) contain many industry links. You can organize a search by products, services, or industries, and then use the results to move in any direction you choose.

Deciding on Graphics

To this point (this is an example of summarizing) we have examined approaches for conveying your information, the importance of purpose and context, possible organizational schemes and formats, the process of incorporating information from other sources into your own message (synthesis), and sources for researching your topic. Now we move into another set of organizational choices: the graphic presentation of information.

Graphics can take many forms, but they are always used to provide a *visual* representation of information, and ideally they should link your audience with your conclusions. Because they can help your audience grasp and retain critical points, they are not merely window dressing. Below are some graphic formats you can consider. Your choices should be guided by such criteria as what is appropriate given the data, preference of the audience, your presentation format, and personal preferences. In addition, if you are using color monitors, color projection equipment, or color transparencies, you should choose colors that will stand up well to such projection. For example, a yellow background with white type will not show up well, while a dark blue background with white type would. We analyze such issues in more detail in Chapter 6.

TABLES

A table is a systematic arrangement of data in rows and/or columns. Putting information in tabular form makes it more accessible because facts are categorized and arranged systematically. Also, tables make the comparison of several years or types of information easy. There are several types of tables, including *ruled tables* (headings set off with horizontal lines) and *boxed tables* (data separated by both horizontal and vertical lines). Table 4–1 is an example of a boxed table; it includes data that are used in the figures that follow.[2]

FIGURES OR CHARTS

A figure or chart provides overall information and impressions. Figures and tables complement each other. In a report you might present data in a table and highlight trends in a figure. Tables provide the audience with *exact* information, while figures concentrate on the visual impact of trends. At a glance the audience knows if sales have increased or decreased, if stock prices have gone up or down. The audience also gains an impression as to the size of the increase or decrease. Typical

Table 4–1   A Boxed Table

Smith Company Competitive Analysis

	Profit Margins			
	Model 100	Model 200	Model 300	Model 400
Old	12%	5%	28.8%	0%
New	45	47	52	42

	Market Share (millions of $ sales)			
	Model 100	Model 200	Model 300	Model 400
Total market	400	350	600	50
Smith sales	20	21	210	15
Market share	5%	6%	35%	30%
Profit margins—old	12%	5%	28.8%	0%

	Competition (millions of $ sales)			
	Model 100	Model 200	Model 300	Model 400
Smith	20	21	210	15
Jones	89	92	48	2
Blue	125	31	34	3
Apex	25	59	140	5
Others	141	147	168	25
Total market	400	350	600	50

types of figures include line graphs, pie charts, and flowcharts. With various figures, you might be able to choose a three-dimensional option where the figure is presented with shading representing three dimensions.

Line graphs. In line graphs the dependent variable is always shown on the vertical axis (y-axis), the independent variable is always on the horizontal axis (x-axis). Some line graphs use more than one line in order to compare figures (see Figure 4–1).

Bar graphs. Bar graphs also use horizontal and vertical axes, but unlike line graphs, the dependent variable can be shown on either axis. All the bars on a bar graph must be the same width; difference is expressed only in the height of the bar (see Figures 4–2 and 4–3).

Figure 4–1   A line graph

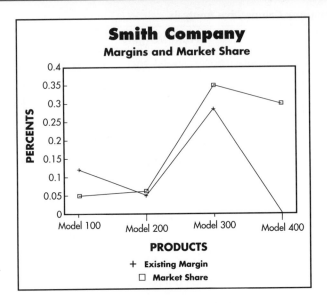

Figure 4–2   A bar graph

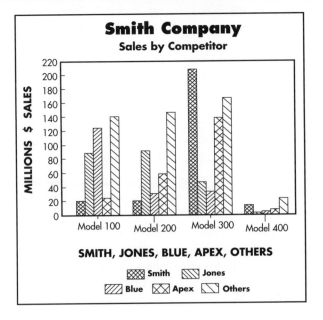

Figure 4–3   A bar graph

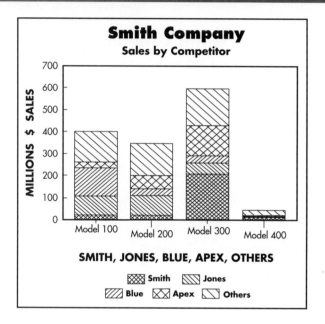

Pie charts. Pie charts are used to show percentages of a whole. The pie chart represents 100 percent and is divided into percentage segments. Each segment of the pie should be labeled; crosshatching or shading can be used to set off various segments (see Figures 4–4 and 4–5).

Flowcharts. Flowcharts are used to show processes. They can be used to illustrate sequential steps or to show the relationship between departments or units.

COMPUTER GRAPHICS

New technology allows presenters to choose from a whole array of computer-generated images. Computer graphics allow you to be more flexible in your development and presentation of visual aids. You can experiment with size and format as well as with placement of graphics on the page. The computer does all the cutting and pasting that used to be done by hand. There are many types of computer graphics programs available; the following are examples:

Spreadsheets. Spreadsheets assist in forecasting, planning alternatives, analyzing trends, graphing results, and producing reports for decision making. A spreadsheet program also can create graphs based on the information in the spreadsheet. However, the graphic potential of some spreadsheet programs may be too limited for sophisticated presentations.

Figure 4–4   A pie chart

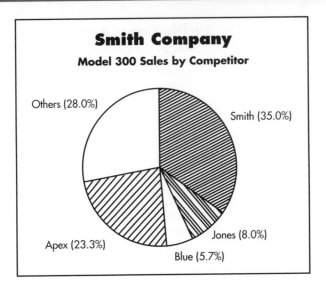

Figure 4–5   A pie chart

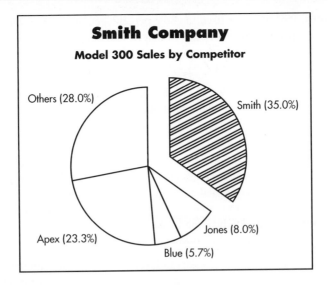

Dedicated graphics packages. These software programs provide more options and, in most cases, better quality visuals than those created by spreadsheet programs. Most of these packages allow you to import data from spreadsheets, eliminating the tedious step of reentering data.

Integrated software packages. These programs combine word processing, spreadsheet, and graphic capabilities into one package. They offer convenience to the user and allow easy access to data and formats.

Presentation software. Certain software packages allow you to create transparencies containing text and graphics. In addition to creating overhead transparencies or computer-generated transparencies for projection as part of an oral presentation, these programs allow you to have speaker notes and handouts for your audience. These programs are especially useful for oral presentations but, in some circumstances, can also be used to create useful tables and figures for written reports.

Internet graphics. Internet sources include many copyright-free graphics. In addition, when working for a client company, home pages often include logos, photographs, and other company material that can be used with permission.

Summary

No communication plan is complete until you have thought through an approach and until you have gathered all the necessary research data and considered the scope of your content. After deciding on approach and content, you will have devised a thorough communication plan, consisting of purpose, audience, and content. You are now ready to begin drafting your text or your oral presentation. Remember that as you draft, you may realize that some of your initial decisions about approach or content may need revision. As you create the text, you will probably notice where additional research would help or that your ideas would make more sense in a different framework. The communication plan you have devised is not set in stone; it is merely a tentative formula to guide later choices. Be flexible in its use, and be ready to revise as the communication context warrants.

How to Approach Practice Tasks

ANALYSIS VERSUS DESCRIPTION

1. Three facts (description) are:
   a. Between 1959 and 1971, average cruising speed remained about 600 mph.
   b. This information starts with the advent of jet airplanes and excludes older propeller-driven ones.
   c. Capacity really increased with the introduction of jumbo jets in 1970.

2. Interested parties could be:
   a. Airline history buffs.
   b. Students looking at the transportation industry.
   c. Analysts looking at the evolution of the airline industry.
3. Analysis for the three interested parties might include the following:
   a. Airline history buffs: While Boeing and Douglas (which later merged with McDonnell and then merged with Boeing) dominated aircraft chosen by United in the early jet era, the Caravelle and Viscount played an interesting role in short hauls. These aircraft were generally liked by both business and vacation travelers because of their spacious interior and unique furnishings.
   b. Students: Right from the start, what is now the modern Boeing Company was dominant in equipment choice. The Caravelle and Viscount proved to be aberrations. These early successes led to Boeing's ability to provide capital for the development of later series (e.g., 767, 777) and the eventual elimination of McDonnell-Douglas as a competitor in the commercial market. In turn, this led to Boeing's acquisition of McDonnell-Douglas as a way to control both the military and commercial aviation fields.
   c. Analysts: In today's world, Boeing and Airbus are virtually unchallenged as leaders in producing commercial aircraft. This table shows how the groundwork was laid from the beginning of the jet era. From 1960 onward, Boeing (with some minor digressions) was the supplier of choice of major airlines such as United. The advent of the jumbo jet in 1970 provided the beginning of a very profitable period for both Boeing and its customers.

FORMAL VERSUS INFORMAL APPROACH

Perhaps the most formal approach will be when you are a professional witness. Specific and formal language is required so that there is no misunderstanding of your testimony. Any use of jargon or unfamiliar terms is accompanied by definitions. Your tone and demeanor will be part of your credibility as a witness. At the other end of the spectrum would be when you speak to government accountants. Here you can be informal and use jargon familiar to your audience. In the middle is your presentation to the board of directors. Formality might be appropriate in some companies, while in others you are expected to be quite informal in tone and approach.

DIRECT VERSUS INDIRECT APPROACH

- **Direct:** Our team must conclude that Pat Simpson (vice president of finance) is not doing an adequate job of providing information to management. Because of this lack of communication, management decisions are being made without the proper background. Several recent decisions have resulted in misunderstandings with shareholders.

- **Indirect:** Pat Simpson (vice president of finance) is responsible for the generation of external reports to shareholders and taxing authorities. Our team has found that he is performing these functions well. However, there is a problem in relation to Simpson's overall communication with management.

SUMMARIZING

The main idea is to check with your audience to see if you have properly summarized. You might first ask the audience to read/hear your summary. Then, provide a copy of the article and ask the question again. You want to know if you left out salient points or spent too much time on points that were tangential to the main thrust of the article.

SYNTHESIS

At the time of the introduction of FAS 128, some writers stated that the new rule would make earnings per share look higher for companies that had simple capital structures as compared to those with more complex structures. Others argued that investors would see through this as just a new way to state results: Nothing had really changed fundamentally about the company. The synthesis could not only deal with these contrasting views but also include a more recent analysis of whether the concerns of these writers did or did not come true.

Endnotes

1. J. K. Thomas and S. Tung, "Cost Manipulations Incentives under Cost Reimbursement: Pension Costs for Defense Contractors," *The Accounting Review* 67, no. 4 (October 1992) p. 693.

2. Data in this example adapted from Dagmar Bottenbruch and Robin Cooper, *Mueller-Lemkuhl, GmbH* (Boston: Harvard Business School, 1986).

Part 2

Implementing a Communication Plan

> 66 Public accounting requires its practitioners to be able to transfer and receive information with ease. . . . Practitioners must be able to present and defend their views through formal and informal, written and oral, presentation. They must be able to do so at a peer level with business executives. 99

BIG EIGHT STATEMENT ON ACCOUNTING EDUCATION

Chapter 5
Creating a Text

You sit at your desk, with papers and folders arrayed in front of you. The clock is moving toward midafternoon, and everyone else in the office looks busy and focused on important tasks. Your computer has a completely empty screen, and you desperately hope that no one walks by to see that you're still struggling with the feasibility study for Midwestern Industries, the pet project you begged for and the job that can really show what you can do. All the numbers are crunched, and all the research is done. The report must be submitted by tomorrow afternoon, and it will be, if you can just write it up.

Now is the time to confront your old English 101 demons. Your ideas have always been creative and important, but the papers never seemed to be as clear and focused as they should be. You usually procrastinated until a writing paralysis set in; today is no exception. If you could just get started, you would be OK, you tell yourself. For now, though, you just look inert to your office mates.

To this point, we have asked you to consider much of your accounting experience as a series of communication challenges, but we have also suggested that with thoughtful planning, you can control and direct many aspects of the communication process. Although communication is often complex and unpredictable, it is less intimidating if you acknowledge which parts of the process you can reasonably control and if you develop and follow a reasonable communication plan. In this chapter we will suggest some strategies for drafting, the next stage in the communication process. You will learn ways to turn your communication plan into a text, initiate the drafting process, begin working with your evolving text, and break through writer's block or speaker's apprehension.

From Plan to Text

Successful professional accountants must understand how to turn a communication plan into a coherent message—a text—to which an audience will respond. Several factors are important to remember as you create texts for written or oral communication.

1. Remember that having a specific communication plan does not necessarily mean that you have to stick slavishly to that exact plan. In public speaking and in writing, the prime purpose of planning is to provide a foundation that will encourage you to be flexible and spontaneous as a speaker or writer.

2. Recall the main point in Chapters 2 and 3: Planning a message must include a clear understanding of the audience for that message. Time spent analyzing your audience will probably result in a clearer sense of what you want to say and how best to say it. Taking the role of potential readers and listeners, getting outside yourself, is the surest way to judge whether your plan will be effective before it actually reaches the audience.

3. Assure yourself that the more attention you give in the early stages to preparing a text, the more confident you will be in presenting your message. It is important to do your homework and consider all your alternatives carefully.

4. Understand, as discussed in Chapter 4, that the order in which you present your ideas will make a difference to most audiences. Receivers accept and remember some pieces of information primarily because of where they are placed in an overall presentation and, conversely, will tend to forget other ideas placed less strategically. Generally speaking, readers and listeners will be more likely to recall what you say at the beginning of the presentation or at the end. The practical implications are clear: (*a*) A communicator will want to prioritize the parts of a message while planning for its presentation; (*b*) particularly crucial ideas, facts, or data should probably be stated both early

---

FROM PLAN TO TEXT

- Remember that you need not stick slavishly to your original plan.
- Focus on the audience for your message.
- Remember that the more attention you give to the text in its early stages, the more confident you will be in presenting it.
- Consider that the order of your ideas makes a difference.
- Think of drafting as a creative act.

---

and late in the presentation, thus the importance of strong previews and summaries; (*c*) important ideas that attract audience interest should probably lead the presentation, while important ideas that point toward future action should probably conclude it; and (*d*) ideas you expect your audience to remember should not be submerged in the middle of a presentation.

5. Recognize that developing the specifics of a text is not a mechanical process with clean-cut techniques or formulas. Drafting is a creative act, and in some ways even an artistic one. Occasionally, writers and speakers try to force creativity by meeting the task head-on ("I'm going to sit at this keyboard until an idea hits me!"). This strategy rarely works, since ideas are stimulated by other ideas. Fortunately, experienced writers and speakers have discovered some tricks that might work for you, too.

Turning Blank Pages into Drafts

DRAFTING AS EXPERIMENTING

Drafts are the raw material of planned written or oral presentations; in a sense, they are the trial balloons you monitor yourself, your experiments with language that you do not need to show anyone else. You write a draft in order to come up with something better that you are later willing to share. Drafts, almost by definition, are inadequate, fragmentary, preliminary, and, in some cases, just plain bad. Since your purpose in drafting is to develop a text to work with, a clumsy first effort is no problem. You know the draft needs work before it is ready for public consumption. In Chapter 6, we discuss strategies for polishing (revising/editing) drafts into texts. However, at this point the question is, "How can I generate something that I can later make presentable?" Remember that no one will see your early efforts—no one will know how many times you rewrote that first sentence.

DRAFTING ORAL PRESENTATIONS

Obviously, drafts are important in writing memoranda, letters, and reports. Less obviously, they are just as important in the process of developing oral presentations. In most cases, the chief difference is that drafting prose involves word-by-word choices for the entire text, while most oral presentation texts take the form of a complex outline. You will note exceptions where a speech involves a fully prepared text (a news release, the president's State of the Union address, the announcement of how balloting takes place for the Academy Awards). These are contexts where, for legal or organizational reasons, a full-blown text is needed and read. However, in most oral presentations, reading from a text stifles the impact of the message. Thus, speakers carefully develop a set of notes, an outline, or a set of key phrases. A speaker can glance down at notes, see the key point, and then flesh it out as she talks about it. This outline, or set of speaker's notes, forms the basis for the text of the message, which is spoken extemporaneously. As the speaker practices, each session becomes a means of redrafting as full sentences and paragraphs fill out the key points from the set of notes. In some cases, the speaker will want to make sure the notes include very specific words or phrases that are important for the message or for providing transitions. This basic process is started in the same way that a text writer uses an outline, key points, and specific phrases as he sits down to draft a full text for a report.

INVENTING A DRAFT: GETTING STARTED

Too many people try to produce final copy at the draft stage and end up paralyzing their critical and creative abilities. They believe they can save time by short-circuiting the planning process and progressing directly to the product they want to communicate.

When you get that stuck feeling (sometimes called *writer's block* and experienced by many people, even professional writers), you could quit. However, remember that until you get something down on paper, no matter what it sounds like or looks like the first time, you have nothing to work with. Keep reminding yourself that this is only a draft; it doesn't have to be perfect.

Here are suggestions to help you get your creative juices flowing and your first draft produced.

1. **Arrest your inner critics.** Most of us have been socialized to think that some of our ideas are wrong, stupid, or socially unacceptable. "That'll never work," or "That's not what they want," says the critic of our inner dialogue when we first have the glimmer of an idea. Of course, many ideas ultimately will not work. However, no idea that is dismissed immediately can develop its own life. Hunches, clues, and tentative insights are ignored and left to wither. Parents do not look at infants and say, "What good is this helpless creature?" Instead, they assume the baby needs time and nurturing in order to grow. Similarly, writers and speakers need to let their ideas breathe at first by not smothering them with criticism too soon.

2. **List all relevant ideas by brainstorming.** *Brainstorming* is often used as a group problem-solving technique in business, industry, and education. Whether done in a group or by an individual, the goal of brainstorming is to promote creativity and to keep it from being hampered by assumed limitations. It emphasizes idea-generation over idea-evaluation. Within a time limit, brainstormers generate as many alternative ideas as possible.

While creating a brainstorming list, take the word *list* literally; a list is a collection of things in no particular order. At the getting-started stage, do not worry about how unrelated ideas might ultimately hang together. Just get them all down on paper. Do not prejudge the ideas; record outrageous, ridiculous, or unlikely notions along with the ones you consider immediately reasonable or possible. Just record anything that comes to mind as a result of your analysis and fact gathering. No criticism is allowed at this initial stage. You might want to consider doing your brainstorming in several sessions, when you are in different moods, in order to tap different parts of your experience.

3. **Cluster the items on your list according to similarities of topic, data, style, and audience interest.** Look at your list and find the items that seem related to one another; you will find items that have things in common. By clustering these items together and then moving clusters around, you can both create the basis for full text as well as see how your ideas fit with your communication plan. This process then leads into editing and evaluating your list.

4. **Begin to evaluate your ideas.** You started off with a statement of purpose and a communication plan. See how the items on your list accomplish the purpose and plan. Also consider how the purpose and plan should be altered given your list. View items for clarity, good sense, appropriateness, and persuasive appeal. Given page or time limitations, some ideas might have to be dropped. While important, they need to be edited so that the main points of what you want to say are stressed. However, before you cross anything from your list, ask yourself if there are parts of that item that might be salvageable and used elsewhere or in another way. Sometimes important seeds of creativity are hidden in an item you might eliminate as being an impossible idea or an impractical way of looking at things.

5. **Arrange your clusters of ideas into a tentative organization.** Chapter 4 suggests several possible organizational schemes. These strategies are common and practical ways for presenting information, analysis, implications, and conclusions or recommendations. As you arrange your clusters of ideas, also consider how you will use transitions to move from one

---

PRODUCING A CREATIVE DRAFT

- Arrest your inner critics.
- List relevant ideas by brainstorming.
- Cluster items by topic, style, and audience interest.
- Evaluate ideas.
- Arrange ideas into a tentative organization.

Brainstorming

You have been asked to make a presentation to upper management about the company's plans to acquire a new subsidiary in a related line of business. You have at your disposal the history of the new division, including both financial and nonfinancial information, and the analysis that your superior used (and you assisted with) to convince the president that the acquisition was a good idea. You are having trouble getting started.

Tasks:

1. How would brainstorming differ if your presentation is to be in writing rather than a formal oral presentation?
2. Choose either format above and brainstorm about possible choices for the presentation.
3. Group the ideas you come up with.
4. Filter good ideas from those that do not seem to be workable.
5. Make a list of questions that you want answered before you can start a draft of your presentation.
6. Which organizational framework would you use?

---

idea to the next in the body of your presentation. Create some smaller lists where you develop how you will (*a*) introduce your presentation in a way that clearly states your purpose and catches the audience's interest and (*b*) conclude by linking back to your purpose, summarizing, and pointing the audience to the next step that should be taken.

Strategies for Written Communication

People who write daily in their jobs usually develop other little tricks that help them through periods of writer's block or when creativity just does not seem to flow. Here are some basic methods for getting started that have appealed to writers over the years.

Two reminders are in order. First, these methods must be understood in the context of drafting-as-experimenting, developed earlier in this chapter. This attitude is more than a technique; it is an overall appreciation that ideas come from other ideas and that if an external conversation is not immediately available, you should invent an inner conversation to stimulate the thought process. This section describes straightforward strategies or techniques that will reinforce your basic appreciation of writing as a process of experimenting and creating for a particular audience. Without this attitude, the techniques are meaningless. The second reminder is perhaps more obvious: These strategies for written communication will also apply to oral communication situations, although we later offer ideas specifically tailored for your speaking responsibilities.

Consider these possibilities for increasing your creativity in drafting text:

FREEWRITING

When people complain of writer's block, writing is not actually blocked in any real sense; instead, ideas feel blocked or, in some cases, the writer's mind feels totally empty. This view presumes that the mind should be like a warehouse of ideas. The ideas are stored in anticipation of a proper transportation system, perhaps a conveyer belt, to move them onto the loading dock of a year-end report or memo to the marketing department. With writer's block, though, it's like the warehouse office receives an order for merchandise that just isn't on the shelves.

In contrast, some authors experience writing differently and use different metaphors. Poet and essayist William Stafford, for instance, sees writing as more similar to swimming than to ordering from warehouses or tapping reservoirs of ideas. Stafford, in his book *Writing the Australian Crawl,* points out that in water the swimmer finds no handholds and quickly discovers that to stay still is to sink. Instead, it is only through movement and proper activity that someone can stay afloat and finally get somewhere. Ideas present similar challenges. Writers get them only by starting to write. Without actually starting, you have nothing to suggest where the next ideas will come from. You sink. Stafford is right: Water (and writing) has no handholds or guarantees, so you stay afloat by trusting the movement itself.[1]

Effective writers begin the process, then, by starting to write. One idea-generating process taught in some classes is *freewriting.* The basic rules of freewriting are to set a time limit (10, 20, or 30 minutes are usually enough) for the writing, and don't stop writing until the time is up. If you cannot think of what to say, write "I can't think of what to say" and continue to write about how you feel about not having something to say. Although freewriting in classes is often assigned without a specific topic focus, the basic method will also work for you if you already know the topic you want to explore. Freewriting gives you a draft, though it's very rough and unrefined.

AUDIOTAPING

For some writers, the blank paper or blank computer screen can be intensely discouraging because the empty space symbolizes a failure to come up with ideas.

BREAKING FREE FROM WRITER'S BLOCK

Try this idea the next time you encounter writer's block: *Don't start at the beginning.* Many people worry and fuss over the first sentence of their texts, essentially blocking the flow of words until they get the opening "right." This behavior is counterproductive and a waste of time. If you are having trouble getting started, begin working on a portion of the text you feel more comfortable with—the middle section or even the conclusion. Writing is like priming a pump; once you get started, the text will begin to flow and the block will dissolve. No one will know where you began drafting; your readers will see only the final product.

Freewriting

Open an accounting book to a section you have studied. Pick a subject from the chapter that seems difficult for you to understand. Review what the book has to say on it. Then close the book and freewrite for 10 minutes on this subject.

The blankness is not an invitation but an intimidation. If you find it hard to generate text on blank paper or monitors, you might consider audiotaping your first draft. Use a cassette recorder in a private place where you will not be interrupted and no one will be able to hear you. If you have already developed a tentative outline for your report or other writing project, speak extemporaneously directly from the outline, recording your ideas informally. If you have no outline prepared, consider your taping as a form of oral freewriting—freespeaking. Mixed in with your idea nuggets will be some fool's gold, and you are likely to record many speech mistakes as well. Yet when you later sit down to transcribe what you said, you will have a draft and no blank pages to worry about.

Another audiotape strategy that you might try is to tape a conversation you have with a close friend or colleague about the topic of your writing project. Although this process will not give you a draft as directly as taping your freespeaking, it will help you keep in mind the important ideas that you generate with another knowledgeable person. Additionally, recording a focused conversation may give you good preliminary clues about how an audience might react to the ideas that you eventually use in your text.

IGNORING COSMETIC OR SURFACE CONCERNS

Drafting a text is not the time to polish it (see Chapter 6). Therefore, forget about spelling, punctuation, grammar, and even neatness while putting your draft together. You can read your own writing, and you will understand what you mean by certain verbal shortcuts ("cd" for "could," "w/ot" for "without," "com" for "communication," etc.). Do not start counting your words yet or try to plan what kinds of margins, fonts, paper, or envelope your final report will sport. When you print it out in draft form, at least double-space your text to invite further editing. Single-spaced text on a page or screen may cause you to overlook opportunities for improvement while editing and revising.

Strategies for Oral Communication

Public speaking is an activity that most people fear and try to avoid in their everyday lives. You often feel your reputation is at stake when you appear before others. The public speaking experience can be intensely involving for both speaker and listeners, and speakers often feel as if they put their egos on the line when they talk in public. Criticism is difficult to hear most of the time, but for some rea-son, criticism of how a person speaks is often heard as criticism of who the person is. This puts great pressure on some speakers. Chapter 8 presents ways to deal with the apprehension many people feel in speech contexts.

At the same time, most executives who have been around for a while observe that those who are willing to speak in front of others receive extra recognition and rewards within the organization. The willingness to speak implies confidence, leadership, preparation, and credibility, as well as a sense of audience needs.

Tips on how to draft a speech text must take into account the extra complica-tions of recognition in organizational life. Early in the planning and drafting stages, prospective speakers must decide what kind of speaking style helps them feel most comfortable and what kind of role (persona) is appropriate for the anticipated tasks.

Therefore, in addition to the strategies already presented for written drafts, consider these additional suggestions for the early process of planning for oral presentations:

LIMIT THE PRESSURE

Since everyone is open to evaluation and criticism within public situations, decide early to focus on those things you can affect and change, rather than those over which you have no control. If you are sensitive about being overweight, bald, too tall, too short, or too whatever, that's OK, but these things really don't have much to do with preparing your speech. However, if you worry persistently about them, they may prevent you from improving the very parts of the presentation that will impress an audience no matter what you look like.

Successful public speakers have a solid sense of who they are and do not have a great need to portray a different image. One piece of advice beginning public speakers have found helpful at times is "Don't wow 'em." This means that with few exceptions, you should speak from within the comfort zone of your own per-sonality. Be yourself. You should not strain to emulate famous speakers, to be more charismatic than you feel, or try to adopt a show business air if you are uncomfort-able doing so. After deciding to make a public presentation, do a small personality assessment by asking: "In which social situations have I felt most comfortable?", "In which social situations have I felt most phony and artificial?", and "How have I talked in the past when my colleagues have listened carefully to me?"

PRACTICE EXTEMPORANEOUS SPEAKING

As you learned earlier, extemporaneous speaking is the presentation of carefully prepared ideas in the style of spontaneous speech. It is the midpoint between total spontaneity and total planning, between winging it and spouting canned, pre-processed ideas. Extemporaneous speaking relays two important messages to an audience: that the speaker cares enough about the topic to be prepared, and that he or she cares enough about the audience to be flexible. If listeners appear confused, or even especially intrigued by one topic in the presentation, the extemporaneous

> ### TIPS FOR GETTING STARTED IN ORAL PRESENTATIONS
>
> • Limit the pressure—be yourself, determine how to make yourself the most comfortable.
> • Practice alone and in front of others.
> • Adopt a conversational or storytelling style.

speaker is confident enough to abandon the outline and make an explanatory side trip. This skill is not automatic. It comes with hard work, careful planning, and much practice.

To practice extemporaneous speaking by yourself, speak from your preliminary idea list or from your tentative outline—but speak in private where no one will hear you. Speak as if you had the audience you will eventually address. Do not interrupt yourself, but listen carefully to the word choices you tend to make each time you run through the speech. (You may decide to tape your practice sessions, but if you do, be sure you have time to review the tapes carefully. With limited time, your best strategy might be to practice a few extra trial runs of the speech.) Much as your written drafts can suggest new and surprising ideas to you, your extemporaneous practicing will, in effect, teach you what is most important about your message and which words might be jarring or not quite right in advancing that message.

It's also a good idea to practice in front of others as the next step. From a content perspective, this allows you to see whether your intended meaning is really getting across to your audience. Also, many of the unconscious things we do (reading too much, fumbling with notes, saying "uh" or "you know," dropping our voices at the end of a sentence so it's hard to hear, or standing with our hands in our pockets) can be picked up by a friend who is watching. You can correct these behaviors before you are in front of your real audience.

ADOPT A STORYTELLING (CONVERSATIONAL) STYLE

Most oral presentations involve a conversation in some sense. While much of the presentation is one-way, your audience is reacting to you as you speak. You are not there to entertain; you are there to deliver a message in a form that is most comfortable to both the audience and you. Think about a recent conversation among your friends or co-workers. Perhaps a colleague, Karen, is complaining about the politics of developing the most recent budget. Chances are, her speech is characterized by a conversational tone, animated gestures, natural movement, insistent eye contact with listeners, a tone of voice that emphasizes the important parts of the message, and moments of silence inserted into the story probably for suspense. She slows or alters her tale a bit if someone looks at her quizzically; she brings them back into the story with a few extra details. If she were asked to give a presentation at the next company meeting, she might reject the opportunity because she has no experience in public speaking. Yet she has the best preparation

Experiences in Public Speaking

Talk with three professional accountants about oral presentations they have given. At first, do not discuss with them the ideas of this chapter or share with them your own experiences in public speaking.

Tasks:

1. Ask if they are satisfied with their success at presenting detailed information persuasively so that audiences will listen to and remember the main points.

2. Find out how they decided what to and how to organize their information.

3. Do they seem to adopt an extemporaneous, storytelling style while presenting information precisely and clearly? Why or why not?

4. Do the answers to Tasks 2 and 3 shed any light on the answers to Task 1?

for involving an audience—if she only knew how to focus it. She (and you) should realize that there is no mystical formula for public speaking; it is simply the process of involving an audience in an important message that is expressed in her (your) own natural conversational or storytelling style.

Summary

Because it is an essential and often frustrating process, creating a text can be the hardest stage for a communicator to complete. It involves a number of psychological, as well as technical, factors. Although it sounds odd to think this way, creativity is often a matter of roaming freely through the range of possible ideas for any given subject and then sneaking up on yourself, in a sense, to discover what you might eventually want to say to your audience.

In this chapter we see how brainstorming, freewriting, stating ideas aloud, and practicing speaking are ways to bridge the gap between a communication plan and creating a draft on paper. With ideas on paper, you can begin the process of editing, clarifying, and redrafting.

How to Approach Practice Tasks

BRAINSTORMING

In this case, we do not have a sufficient amount of information to do the whole job, but we can see how brainstorming helps. In brainstorming, your first goal is to generate as many ideas as possible and to forget about evaluating them at this stage. You have done basic research on the proposed acquisition but may not have fully decided on relevant assumptions or how to fit information together. Do not worry about censoring your assumptions or whether things are in the right order or connected. Just get ideas on paper. At this stage, whether the final report is written or oral is immaterial as well; you are developing a list that will later turn into an outline and text.

In your brainstorming, you might also get some ideas about presentation format. Perhaps a certain idea is best shown as a graph rather than as a text paragraph (or just an oral explanation). Looking at your list might remind you of some parts of the presentation that could be in number or bullet list form rather than in paragraph form. Ideas for major and minor headings might also become apparent.

Now, let's look at some of the information you have at hand. You have at your disposal a good array of financial and nonfinancial information. What are the most important parts of that information that you want to present? Some brainstorming ideas might include the following as a small sample:

- I want to tell them about how this new subsidiary came into being; there's an interest story there.
- Financial history—it's strong and a good selling point.
- How about a metaphor to organize the presentation? I could use a baseball metaphor in which this new subsidiary is a needed addition to the team. Perhaps it's more like a good meal where this is just the right ingredient to make the whole recipe work together.
- Tell them about the people who work on the production floor—a loyal workforce with an average of over 15 years' experience.

After you think you are finished with your brainstorming, before you begin to cluster ideas, ask yourself if you silently censored any ideas as being weird, strange, or unworkable. If so, add these to the list at this point. Remember that you do not have to play it safe when brainstorming; anything goes at this point. You want to promote rather than limit your creativity.

The metaphor idea can be a good one to help group ideas under an organizing framework. In addition, a metaphor allows a different way to organize a presentation along topical lines.

The rest of the process really relies on how far you have come to this point. If you go through Tasks 1–3 carefully, then take the time to go through filtering ideas, asking yourself questions, and deciding on your organizational framework.

FREEWRITING

This is like brainstorming in many ways. The idea is just to put down on paper whatever comes into your mind. At this point, write anything you think is germane to the subject. Do not filter whether it's right or wrong. That can come later as you compare what you have written to the text and class notes. This is not a test; it's just a way to get you to put some of your thoughts on paper for your own future clarification and review.

EXPERIENCES IN PUBLIC SPEAKING

Try to pick people who are not professional speakers. Many new accountants join organizations like Toastmasters where they are asked to practice speaking. Thus, you might talk to them. It's also useful to discuss various levels of oral presentations. Some are quite informal with everyone sitting around a table—no slides or graphics. At the other end of the scale are fully orchestrated formal presentations with a podium and extensive use of visual aids. What is the same and what is different for these extremes among those you interview?

Endnote

1. W. Stafford, *Writing the Australian Crawl* (Ann Arbor: University of Michigan Press, 1978), pp. 21–28.

Chapter 6

Polishing Texts and Presentations

Imagine you have just taken your seat for a day-long orientation session for your new company. You were recently hired, and today is your first full day on the job. You are looking forward to learning more about your new work environment. The first speaker introduces herself and begins her PowerPoint presentation. The first slide comes up; the company logo is superimposed on a picture of the office building. The caption reads: "Welcome to Frankford Industries—Your Important to Us." At first you are impressed with the graphics, but then your eye is drawn to the word "Your." It's the wrong word. The presenter does not seem to be aware of the problem. You cannot seem to see anything else. You wonder how many other mistakes will show up on the slides. More important, you begin to wonder about the polish and professionalism of your new company.

This presenter has made a classic error; she forgot to proofread her material. Once off on the wrong foot, she may have trouble winning back the confidence of her audience. As a final step in the communication process, your written message needs to be revised and proofread thoroughly, and your oral presentations should be practiced and polished. In this chapter you develop strategies to help you be a more effective reviser and proofreader, understand the role and limitations of using computers in the revising/proofreading process, learn to avoid the most common mistakes in English grammar and usage, learn how to assess the final page layout, and develop practical strategies for polishing an oral presentation.

The Process of Revision

Revision means to look at your draft with *new eyes*. Read your text as though it were written by someone else—a process that is more easily described than done.

Writers often have trouble being objective and critical of their own writing. It's not that we think we are wonderful communicators, but we know what we wanted to say, and we readily assume the message has been conveyed by the text we have created. Often we do not discover a problem until we get an unexpected reaction from our audience. The audit procedure that you carefully described is misunderstood by your staff; the memo you meant to be humorous causes some of the staff to take offense. When your audience cannot understand what you have said or reacts in a way you did not intend, you have probably not done a thorough and objective job of revising.

Several strategies will help you revise objectively:

1. **Let some time pass between drafting and revising.** If possible, begin to revise a day or more after drafting your text. Of course, in the crunch of deadlines and work schedules you may not have the luxury of this kind of time. At the very least, before you tackle revision, get up from your desk and take some time away from the task; you might get a snack or take a walk. When you return, you will be fresher and more objective.

2. **Imagine yourself as your intended audience.** Read what you have written from their perspective. Look for sentences that may lack clarity; check the organization of your paragraphs to ensure your ideas have been presented logically; see that there are appropriate transitions from one section to the next. Consider whether your message is clearly communicated. For example, if your purpose is to suggest the changes that a client must make in response to a new Financial Accounting Standard, look at your message through your readers' eyes. If you suspect your readers will have to struggle to understand your meaning, this is your signal to revise. In most situations, you want the audience to be able to follow your train of thought, from first sentence to last, without having to reread any sentence more than once.

3. **Read your text out loud.** This allows you to listen for awkward phrases or sentences that are too long and convoluted to read in one breath.

4. **Reread the entire text from start to finish from the audience's point of view.**

Consult your communication plan often as you revise, to ensure that the text fulfills your goals, meets the needs of your audience, and contains all the information you intended. If you see problem areas, based on your communication plan, adjust your draft accordingly.

Eliminating Clutter

Many accountants switch to a forced, stiff style when they write on the job. Instead of writing sentences that flow smoothly and naturally, they create sentences with strange words and unusual word orders. Phrases such as "enclosed please find" are used instead of simply saying "I have enclosed." If you were talking to someone, you would never tell them "enclosed please find." This time-worn and ineffective business writing style puts a wall between you and your reader. Since your goal is clarity, see if you can remove *clutter*. Clutter can take the form of using more words than are necessary ("due to the fact that" instead of "because") or throwing in phrases that add nothing to the communication or that bear little resemblance to conversational English. For example, "allow me to say that" is literally asking a reader's permission to write what you are going to be writing anyway, with or without permission. A phrase like this is unnecessary. Writers and speakers often introduce clutter by trying to avoid using "I" or "we." In many business communications, using the first person is appropriate. Here are some examples of typical business clutter and some suggested alternatives.

"Cluttered" Expression	**Alternative**
acknowledge receipt of	we received
allow me to say that	(omit)
as the case may be	(omit)
at the present time	now
attached please find	I have included
based on the fact that	because
due to the fact that	because; since
enclosed please find	I have enclosed; enclosed is
herewith	(omit)
in accordance with your request	as you requested
with regard to	concerning
in the event that	if
permit me to say	(omit)
please be advised	(omit)
prior to	before
pursuant to your inquiry	in answer to your question
thanking you in advance	thank you
we are in receipt of	we received
with the exception of	except for

Note, however, that in some of these there are nuance, tone, and connotation differences that could also be factors. For example, "beg to differ" might not always be considered clutter. It is occasionally simply a less direct, and possibly tongue-in-cheek, alternative to the more confrontational "I disagree."

Revising with a Word Processing Program

Revising a draft is much simpler with a word processing program. You can move text around easily, change the order of sentences and paragraphs, and experiment with alternative phrasings as compared to working with a printed or longhand copy. If you do not like an earlier change, you can quickly delete it and return to the previous version.

As handy as a word processing program is, it presents a possible pitfall for the unwary writer. Documents created in this way look too good, too fast. On screen, writing has the potential to look neat, tidy, and reader-ready long before its time. Printed versions of first drafts also tend to look finished, especially if you print them on a laser printer. Remember that even though that printout looks polished, it's still just a draft and needs your attention.

The Importance of Proofreading

Once you have revised your text, you are ready for the final stage, proofreading or editing. Proofreading is the quality control of written communication. Even if you have written a concise, informative, and well-organized report, one spelling error or one slip in subject/verb agreement can undermine the total effect of your work. The final version of your text, the one read by your audience, reflects on you and on the company or institution you represent. Therefore, no text should leave your hands until you are completely satisfied with its quality. This means that no text should leave your hands without being thoroughly proofread.

Proofreading is a process that requires time, alertness, and patience. Too often, a writer just gets tired of reading and rereading the same text. You can get too close to it and lose perspective. Thus, like revising, you need to build in time to remove yourself from the text before you come back to proofread. Your credibility is on the line. This is the last chance you have to detect mistakes and to avoid any embarrassment. Do not rush through this process; take your time and be thorough.

Strategy for Effective Proofreading

The most effective way to proofread any document is a three-step method in which you go through the text three times, with different goals each time through. Because you don't have to look for everything at once, this method detects more errors and produces more improvements than a once-over proofing; the task is broken into manageable parts. You can concentrate on specific items in each stage, and you will naturally be more thorough.

THE FIRST READING: CLARITY AND SENSE

The goal of the first reading is to focus on the sense of the text. If you are using a computer, be sure to run the spell check or any grammar checking program and save the changes *before* beginning the first reading. Although some people are able to proofread on-screen, most people are better proofreaders if they work with hard copy. Therefore, print out the latest version of your document and do the following:

1. Read through the entire text slowly, checking for clarity and sense. If possible, read the text aloud.
2. Look for word omissions and obvious errors in word usage and punctuation.
3. Use a dictionary to check the spelling or meaning of every word you have doubts about.

4. Verify the accuracy of any numbers that appear in the text. (If possible, use a team approach to check any figures and columns of numbers. You will increase efficiency if one person reads off the numbers while the other is checking the document.)

5. Mark all changes clearly on your text.

THE SECOND READING: LINE-BY-LINE ACCURACY

The second reading is hardly a reading at all in the normal sense. Your goal is to stop yourself from reading the flow of the text and instead look critically at individual lines and words. Perhaps the most common proofing problem is that our eyes move so rapidly from idea to idea that we often mentally supply words that are missing or skip over errors. Instead, in the second reading you want to slow down your eyes and do the following:

1. Use a ruler or a sheet of paper to isolate one line of text at a time, beginning at the top of your document.

2. Do not forget to check every line in the inside address and greeting of letters, as well as the headings of memos and the wording in exhibits and graphics. Many people proofread only the body of their texts, but errors can show up in *any* line of text.

3. Examine each line thoroughly before going on to the next.

4. Use a dictionary to check the accuracy of any word divisions by hyphenation.

5. Verify the accuracy and completeness of bibliographic references.

6. Mark all errors clearly on the text.

7. Make all the necessary changes on your text, correcting all errors found in both the first and second readings. Save the changes and print out a new copy of your document.

Some proofreaders start from the end and work toward the beginning of the document. This allows a line-by-line reading without being caught up in the flow of the ideas in the text.

THE FINAL READING: VISUAL IMPRESSION

The last time through your text, you will be performing the final checks and assessing the overall appearance of the document.

1. Using the copy you marked during the first and second readings, check to make certain that all errors have been corrected in the final version.

2. Inspect each page, checking to make sure that the text is placed appropriately on the page and that there are no errors in spacing (e.g., a heading falls as the last line of a page with the text beginning on the next page). At this stage you may also notice inconsistent spacing (e.g., sometimes a double space between sections, sometimes a triple space) and

---

STRATEGY FOR EFFECTIVE
PROOFREADING—THE THREE-STEP METHOD

• Read for clarity and sense.
• Read again for line-by-line accuracy.
• Read a final time for overall visual impression.

---

inconsistent headings (e.g., mixing side heads and centered heads at the same level of organization).

3. Judge the overall quality of the document, including type size and paper stock.

Proofreading with a Computer

Computers can be valuable in the proofreading process, but they cannot take the place of human proofreading. Many people mistakenly think that once they have run their texts through a spell checker or a grammar checker, their writing is ready to be sent out. The computer helps writers check specific aspects of their texts, but you must understand the limitations of these programs and not make the mistake of giving complete proofreading responsibility to the computer. Nothing substitutes for your own alertness in the proofreading process.

SPELLING CHECKERS

For those of us who are poor spellers or a bit inaccurate on the keyboard, the spell checker is a boon. It reassures, saves time, and prevents embarrassing gaffes. Most programs allow for spell checking while you write. Running the spell check is a logical first step in the proofreading process, but be sure to save any changes.

As helpful as the spell check is, be aware of the limitations built into these programs. They cannot alert you to usage errors. Many times writers spell words correctly but have used the wrong word (e.g., *form* and *from*). These errors also happen because many English words sound the same but are spelled differently (homonyms) and many words sound similar but have very different meanings. Words like *there, their,* and *they're* are homonyms, while *affect* and *effect* have close pronunciations but very different meanings. Suppose you create a sentence that reads "There examining the affect of the new cost accounting system." Since all the words are spelled correctly, the spell check will not point out that you have used the wrong *there* or that you have misused the word *affect.* Thus, if you rely solely on the spell check, many word usage errors will slip through. Also be careful to avoid clicking on the wrong word from among the suggested list within the program. Many words look similar; take your time in making your choice.

GRAMMAR CHECKERS

Many programs are now available to check for grammar, usage, and punctuation errors. Current versions of word processing programs include the ability to check grammar as well as spelling. These programs are also helpful tools, but they too have built-in limitations. As long as you are aware of the limitations and do not expect the programs to find all your errors or take the place of your own proof-reading, then you may want to run your texts routinely through these programs.

If you are using grammar checking software, become familiar with the errors it has been programmed to find. Most programs can locate simple subject/verb agreement problems, common word usage errors, and misuse of punctuation. Each year, more sophisticated grammar checking programs become available. However, English is a very complicated language. The rules for grammar and usage are complex; exceptions exist for every rule. In addition, some of these programs are designed to recognize each word as a particular part of speech (noun, verb, pronoun, adjective, adverb, etc.), but in English many words can function as several parts of speech. For example, the word *back* can be a noun, a verb, an adjective, and an adverb. Grammar checkers simply cannot handle all of the complexities of our language.

These programs are most helpful to people who already possess competent grammar skills but who may have trouble with specific aspects. If you know you often have problems with subject/verb agreement, then you might feel more comfortable about the correctness of your texts if you run them through a grammar checker, looking specifically for errors in agreement.

People who lack basic grammar skills or for whom English is a new language are less likely to see improvement in their writing from the grammar checkers. Many times the messages given by the program are merely *clues* to possible problems; the writer has to understand enough basic grammar to make a decision about possible changes. For example, a message that you are using the passive voice may or may not warrant a change.

Avoiding Common Grammar Errors

If you struggle with grammar or if you want to refresh your skills, consider buying one of the many writer's handbooks available, listed in the Appendix to Chapter 1. Although this book cannot cover all aspects of grammar, we will highlight three of the most common problem areas that cause difficulties for many people: (1) subject/verb agreement, (2) pronoun/antecedent agreement, and (3) parallelism.

SUBJECT/VERB AGREEMENT

The rule for subject/verb agreement is a deceptively simple one: Verbs should agree with their subjects in number (singular or plural). A singular subject requires a singular verb; a plural subject requires a plural verb. The key to subject/verb agreement is correctly identifying the subject of the sentence; unfortunately, this is not always easy to do. Sometimes sentences have elements such as *prepositional phrases* (a preposition and its noun or pronoun object) coming between subjects and verbs; sometimes subjects follow verbs; and sometimes sentences have compound subjects that require special attention. Below are some hints for subject/verb agreement:

1. Don't be confused by prepositional phrases or other elements coming between the subject and its verb.

   WRONG:
   An explanation of the figures *are* attached.
   (The subject is *explanation,* not *figures,* which is the object of the preposition *of.*)

   RIGHT:
   An explanation of the figures *is* attached.
   (The singular verb *is* agrees with the singular subject *explanation.* You wouldn't say, "An explanation *are* attached.")

2. Don't be confused by the use of the word *there,* or any other construction in which the subject follows the verb.

   WRONG:
   There *is* twenty people in the office.
   (The subject is *people,* so the verb needs to be plural).

   RIGHT:
   There *are* twenty people in the office.

3. A compound subject joined with *or* or *nor* (*either . . . or, neither . . . nor*) needs a singular verb if both subjects are singular, a plural if both subjects are plural. If one subject is singular and one is plural, make the verb agree with the subject closest to it.

   RIGHT:
   Neither the accountant nor her assistant *is* available.
   (Both subjects are singular, so you need the singular verb *is.*)
   Neither the accountants nor their assistants *are* available.
   (Both subjects are plural, so you need the plural verb *are.*)
   Neither the accountants nor their assistant *is* available.
   (The subject closest to the verb is singular, so you need the singular verb *is.*)
   Neither the accountant nor her assistants *are* available.
   (The subject closest to the verb is plural, so you need the plural verb *are.*)

4. Use singular verbs with these indefinite pronouns: *one, anyone, everyone, anybody, everybody, nobody, each, every, either, neither.*

   WRONG:
   Each of the staff members *are* outstanding.
   (*Each* is the subject, so you need the singular verb *is.*)

   RIGHT:
   Each of the staff members *is* outstanding.

PRONOUN/ANTECEDENT AGREEMENT

The word for which a pronoun stands is its *antecedent;* a personal pronoun takes a singular or plural form to agree with its antecedent in number. This rule, like the one for subject/verb agreement, is simple, but many people still confuse pronoun forms. We often make our grammatical choices based on what "sounds right." Unfortunately, in the case of pronoun/antecedent agreement, the incorrect form often sounds right because we have heard it said incorrectly so often. Incorrect pronoun form is one of the most common errors in both spoken and written English. Here are some suggestions to help you avoid mistakes:

1. To decide whether you need a singular or plural personal pronoun, find the antecedent and determine if it is singular or plural.

2. Use a singular pronoun to refer to a compound antecedent if it is composed of two or more singular words joined by *or, nor, either . . . or, neither . . . nor.*
   WRONG:
   Neither Sally nor Martha will be given new accounts unless *they* improve.
   RIGHT:
   Neither Sally nor Martha will be given new accounts unless *she* improves.
   (*Sally* and *Martha* are part of a compound antecedent, joined with *neither . . . nor;* the pronoun needs to be singular.)

3. Use singular pronouns to refer to these indefinite pronouns and adjectives: *one, anyone, everyone, anybody, everybody, nobody, each, every, either, neither.* (Avoid using the third person masculine pronoun *he* or *his* to refer to antecedents that might include females.)
   WRONG:
   Everybody should maintain files for *their* own clients.
   (*Everybody* is singular, so the plural pronoun is incorrect.)
   Everybody should maintain files for *his* own clients.
   (*His* implies that all people addressed in this sentence are males.)
   RIGHT:
   Everybody should maintain files for *his or her* own clients.
   (This version reflects a gender neutral usage.)

PARALLELISM

Parallelism or *parallel structure* refers to using structurally similar (parallel) constructions to emphasize the similarity of two or more thoughts in a sentence. A sentence that is parallel seems unified and avoids awkward wording. Many people ignore or simply do not understand the importance of parallel structure in their writing. Revising your text to achieve parallelism is an important step to creating a polished and professional document. Parallelism is also essential in creating lists that are part of sentences or that stand alone, perhaps in a bullet list format. Below are some suggestions to help you recognize and achieve parallelism.

1. Use sentence elements with the same structure to connect compound elements that are similar in thought and function.
   AWKWARD: We did half the trip *by train,* and the rest *we flew.*
   PARALLEL: We did half the trip *by train* and the rest *by plane.*

AWKWARD: Steve said that *he was* feeling ill and *could he* leave work early.

PARALLEL: Steve said that he *was* feeling ill and *would like* to leave work early.

(or)

Steve *said that* he was feeling ill and *asked if* he could leave work early.

2. Avoid incomplete parallelism in making a comparison.

   WRONG: Our accounting staff is more experienced than the other company.

   RIGHT: Our accounting staff is more experienced than the other company's.

3. Avoid abrupt and illogical shifts in subject.

   AWKWARD: A team of auditors spent a month studying the tax records, but no errors were found.

   PARALLEL: A team of auditors spent a month studying the tax records but found no errors.

   AWKWARD: Although the spreadsheet looked impressive, the figures were completely inaccurate.

   PARALLEL: Although the spreadsheet looked impressive, it contained completely inaccurate figures.

4. Put all items in a list in the same structural form. For example, each item could begin with an action verb, each item could be a complete sentence, or each item could be a phrase. The key to creating effective lists is consistency. Choose the form you think will work best and maintain that form throughout the list.

   AWKWARD: The duties of the job include meeting with clients, preparing quarterly tax returns, and you have to keep a record of your billable time.

   PARALLEL: The duties of the job include *meeting* with clients, *preparing* quarterly tax returns, and *keeping* a record of billable time.

   AWKWARD: Please note the following changes in your insurance benefit package:

   - The deductible is now $200.00.
   - Orthodontic coverage.
   - Dependents are no longer covered.
   - Added benefits for nonsmokers.

     PARALLEL: Please note the following changes in your insurance benefit package:

   - The deductible is raised to $200.00.
   - Orthodontic coverage is included.

- Dependents are no longer covered.
- Benefits are included for nonsmokers.
(All items in the list are complete sentences.)

(or)

Please note the following changes in your insurance benefit package:
- $200.00 deductible, up from $150.00.
- Orthodontic coverage.
- No dependent coverage.
- Benefits for nonsmokers.

Creating Gender Neutral Texts

Gender neutral language avoids stereotyping people according to sex. Writers who fail to use gender neutral language or who dismiss its importance fail to grasp that the world has changed. According to most style manuals, gender neutral language should be the only acceptable form in any business context. Guidelines for creating gender neutral texts include the following:

1. Do not use the masculine pronouns *he, him, himself,* and *his* unless the reference is clearly male.

   INAPPROPRIATE: Each supervisor is responsible for a weekly report from *his* unit.

   APPROPRIATE: Each supervisor is responsible for a weekly report from *his or her* unit.

   (or)

   All supervisors are responsible for a weekly unit report.
   (By recasting your sentences to use plural forms, you avoid having to use the *he or she* construction, which many people find bothersome.)

2. Alternate the use of masculine and feminine pronouns throughout your text. For example, one sentence in a report might read, "An accountant should emphasize *his* experience in *his* resume." Later in the text you could say, "An accountant also needs to indicate what computer formats *she* knows." This approach, though nonsexist overall, has a disadvantage as well. If a particular excerpt is quoted, it could well *seem* sexist out of context.

3. Avoid beginning letters to an unknown reader with "Dear Sir." When unsure of the sex of your reader, choose "Dear Sir or Madam" or follow your greeting with the name of the position you are addressing, such as "Dear Account Representative."

4. Avoid occupational terms that suggest the positions are held only by men, except when referring to a specific person you know to be male. This suggestion would apply to such terms as chairman (chair or chairperson), congressman (member of Congress), or postman (mail carrier).

Assessing Page Layout

Once you have completed your final proofreading and have a final copy of your text, your last task is to assess the "look" of the text on the page. You want a text that is inviting to the reader. Memos, letters, and reports that are dense (long paragraphs and small type size) do not encourage a reader to stay with them. A professional-looking document includes adequate amounts of *white space* (areas on the page with no type, including side margins, top and bottom margins, and spaces between paragraphs) to aid reading. White space not only improves the appearance of your text but also helps readers by giving a momentary respite. Too much white space makes a text look thin, while too little makes the text look crowded and unreadable.

If you are working with a letterhead and using a computer printer, make sure the date and inside address are not competing with the lettering of the letterhead. If you are using single spacing, double-space between paragraphs to break up the text.

More complicated documents, such as reports and proposals, require additional consideration as you assess the page layout. If you have included graphic elements, make certain they are properly placed and match the text and format of the rest of the document. Charts and graphs need to have enough white space around them to set them off from the rest of the page but not so much space that they appear not to belong with the surrounding text. Make sure to check *where* a

PRACTICE

Revise a Memorandum

Revise the memo on the following page based on the context described below.

Context: Frank James's automobile repair business employs six full-time mechanics. James charges customers 20 percent over his cost for materials and multiplies the wage rate of the mechanics by two in determining the labor charge. Late in 200A James developed the following estimates for the year 200X.

Payroll, mechanics	$103,500
Rent	7,500
Utilities	14,300
Insurance	6,500
Other overhead costs	16,600
Materials for repair work	48,500

James estimates that 85 percent of mechanic time is chargeable to customers.

James has hired you as a consultant. He is an excellent mechanic but knows little about accounting. The objective of the memo to Mr. James is to show him:

a. What profit he should earn in 200X.

b. What multiple of labor he would have to charge (all other data remaining as originally given) to earn a $50,000 profit.

To: Frank James, Owner
From: ABC Accounting
Subject: Profit Analysis

I am writing this to you to discuss the plan that you recently sent me to evaluate for you. I have reviewed the information and have performed a cost-volume-profit analysis (CVP) to determine 200X profit. Profit was determined by the manipulation of the general CVP equation of sales minus variable costs equals the contribution margin, and the contribution margin minus fixed costs equals profit.

Sales is calculated by doubling the 85 percent of mechanic payroll that is chargeable to customers and adding 20 percent over the cost of materials. Variable costs are then subtracted from this sales figure to measure contribution margin. Fixed costs are subtracted from the contribution margin to estimate profit. Sales is calculated by the above stated formula as follows: $(2)(.85)(\$103,500) + \$48,500 + (.20)(\$48,500) = \$234,150$. Variable costs of $166,300 are subtracted from sales of $234,150 to get a contribution margin of $67,850. Fixed costs of $30,600 are subtracted from the contribution margin of $67,850 for a profit of $37,250.

The above estimate of 200X profit is an estimate and should hold true in the relevant range. In short, to ensure that this is an accurate estimate, costs over which you can exercise control should not be changed. Other fluctuations that may occur are expected to be small and within the relevant range. Therefore, the above estimate should not be effected by these changes.

If a $50,000 profit is wished in 200X, a 2.8065 multiple of labor cost would have to be charged. This is calculated using CVP analysis such as the above calculation of profit. Like the calculations above, all figures are the same accept that we do not double 85 percent of mechanic labor that is chargeable to customers. We will replace the unknown variable of X in the place of the 2 for doubling. The equation, given that all the remaining above information remains the same, is as follows: $X(.85)(\$103,500) - \$166,300 - \$30,600 = \$50,000$. When we solve for X, the multiple of labor cost needed to produce a $50,000 profit is 2.8065. This calculation also takes into consideration the assumptions above and the relevant range.

graphic will appear on a page since many word processing programs will move a graph from where you insert it to a later position. If you have created appendixes, make certain they, too, have enough white space and invite the reader's attention. No hard and fast rules govern page layout, but one good strategy is to look for models of attractive page layouts from documents you have seen. Above all, be critical and objective with your final text. If this document came to you, would you find it inviting to read and would you have a positive impression of the writer based on the look of the document? If you can say yes to both these questions, then you are probably ready to send your document.

Polishing Oral Presentations

You have already learned how writing and speaking present different challenges for communicators. Here we will discuss how speakers might best fine-tune their messages and, more important, themselves for particular audiences and contexts.

Writing is to some extent a challenge of preparing a message almost as if it's a product; authors must anticipate possible reactions (such as readers' confusion and feelings of being insulted) and build in strategies to counteract or avoid them. Speaking with an audience is a different kind of challenge. The immediacy of the situation suggests that instead of preparing a message-as-a-product, your main task is to prepare *yourself.* You must not only supply a meaningful message but be responsive and spontaneous in front of people. Contrary to some simplistic public speaking advice, oral presentations involve much more than simply delivering a dramatic talk. Effective speaking demands practice not only in speaking but in listening and sensitivity. If you have attended a presentation, you know how frustrating it is when the speaker answers a question other than what was asked. Here are five general reminders to help you polish your presentation:

TROUBLESHOOT YOUR SEMANTICS (WORD CHOICE)

Semantics is the study of word meanings and their appropriateness in different language contexts. Consider the draft of your speech, usually a detailed outline, and do a thorough semantic analysis for your given audience. Look for terms that might be socially offensive or too technical. Find alternatives. Which jargon or specialized terminology will likely present difficulties for the listener? Accountants may readily understand the technical implications of an "external audit," but does the sales staff understand? Can you substitute more everyday terms that are just as precise? If not, then you can insert a brief definition, almost parenthetically, in your talk. (For instance, "An external audit was suggested in 1998, which means that an accountant outside the firm is hired to review . . . ") Provide yourself with key words and phrases in your notes so that your definition will be clear and accurate. In addition, plan one or two brief illustrations or examples that will clarify the definition; you may not need to use these during the presentation if everyone seems to be following along smoothly, but quizzical looks or blank stares may tell you the opposite is true.

DIVERSIFY YOUR DRY RUNS

Practice the presentation out loud in as many nonthreatening situations as you can. Many people avoid this stage merely because they "feel silly" while talking to themselves or into a cassette or video recorder. Presumably, however, they would also feel silly in public when the talk they have prepared falls flat. It's better to feel silly when no one knows about it. Dry-run practicing can take many forms.

- Talk the speech through several times with a notepad nearby. You will hear yourself creating some transitions between points that seem to work well, while others don't. In taking notes, jot down the wording that sounded effective and identify trouble spots for later repair.

- Tape-record yourself at least once all the way through without stopping, even if you get tangled up in your words or ideas. Use the tape to refine such things as your introduction and conclusion. These crucial spots must be lively and engaging without taking too much time.

- Try the speech out on a live audience who is willing to critique you from the standpoint of the ultimate listeners. Tell the tryout audience that you want honest feedback, not phony reassurance.

COORDINATE YOUR VISUAL AIDS

The face-to-face situation invites speakers to use several different channels of communication. If this is done skillfully, the audience will understand the message more clearly and trust you more completely. Visual aids are one means of communicating with a live audience, but they should not be chosen as an automatic strategy, whatever the situation. Visual aids should be used only as a *complement* to the verbal part of your presentation. They must add value. Your new graphics software may be nice, but do not let its capabilities seduce you into elaborate displays of simple and straightforward ideas or data. Using visual demonstrations to show off your skills may distract an audience and, in some worst-case scenarios, can even suggest that you see your listeners as simplistic and unintelligent.

Visual aids in the form of charts, diagrams, maps, or even three-dimensional models should be used when: (1) complex content must be summarized quickly for an audience, (2) striking trends will have more impact visually than orally, (3) the relationship of various components of a system are better seen than explained in words, or (4) an audience has expressed a preference for visual information.

Once you have decided which information should be reinforced with visuals, you must decide how to coordinate the visual aids with your talk. Should they introduce a problem or statistic, dramatize it, or summarize it? These are questions that cannot be answered beforehand but must be understood within the context of audience analysis. In all likelihood, the polishing stage is the time when such decisions will come together for you. Through feedback and self-analysis you will develop a clearer sense of where your presentation might be murky, confusing, or too detailed and, therefore, need visual aids.

For example, in a presentation you may stress, "It's important to look at our company in comparison to our competitors." That information will produce more audience impact if it is followed by, "To illustrate, look at this chart. I especially want you to notice where we are dominant (point here) and where we are relatively weak." Examine the graphs in Chapter 4 to see how this point can be illustrated.

PRACTICE YOUR QUESTION-AND-ANSWER STYLE

In all but the briefest and most structured presentations, business audiences will expect you to respond to questions, comments, requests for clarification, and even disagreements. Some speakers leave no time for questions or seem to disdain or avoid them, but they pay a high price in credibility and persuasiveness. In fact, the question-and-answer period can be the most effective vehicle for you to make your point. Plan for it, prepare for it, practice it, and, above all, do not forget about it in the excitement of the presentation itself.

Prior to the presentation, try to anticipate likely questions and practice direct, clear, and persuasive answers to them. Better yet, have a listener ask you tough questions and then tape-record and analyze your best responses. Although avoiding and evading questions is usually counterproductive if not futile, some people suggest that trickery is the next best policy. Ignore them. Organizational communication consultant Robert Doolittle writes that he has "been amazed and distressed in recent years to discover formal training programs which teach executives techniques for deflecting, avoiding, or confusing questions in public encounters." One training organization even teaches executives how to "rephrase hostile or difficult questions so that [speakers can] answer a question they would rather have heard—rather than the question asked."[1] Few suggestions could get accountants in trouble faster. Remember that precision and clarity are the hallmarks of your reputation. Here is a three-step strategy for responding to questions:

1. **Perception check.** Restate the question succinctly in your own words, along with any implications that you believe you perceive in it. Although the perception check shouldn't merely echo the original questions, it can serve several valuable purposes. You verify that you understand a question. You prove to the questioner that he or she has been heard accurately. You include any audience members who may not have heard all of the original question. Finally, you establish yourself as fair, competent, and confident in your ability to explain the new point in relation to the previous one.

2. **Clarification.** Respond to the question in order to clarify your overall point. Remember that although one person asked the question, your whole audience hears the answer; address everyone when you respond to the question. Some questions, including many unfriendly ones, give you the opportunity to restate and reinforce your own ideas, but this opportunity will be squandered if listeners perceive your attitude as defensive. Avoid sarcasm such as: "As I said in the beginning to those of you who were listening, . . . " Usually the most helpful attitude is to welcome questions as opportunities for clarification.

3. **Verification.** After the question and its answer, briefly check back with the questioner to verify that the question has been answered adequately. In many instances, it will have been, and you can then solicit other questions. However, some questioners might remain unsatisfied after your clarification and ask for a follow-up. You will usually want to take follow-up questions, but limit your responses to fairly rapid clarification. Otherwise, additional and perhaps equally important questions will be curtailed. One useful approach to a persistent questioner—whether an attacker or a supporter—is to invite the person to talk with you further after the presentation or when the two of you have more time for conversation. Of course, keep your audience in mind: You might react differently if the persistent questioner

was the president of the company compared to someone at your own job level.

CONDENSE AND REFINE YOUR NOTES

Your entire process of polishing the presentation should be calculated to help you decide what you need as support during the actual presentation. Notes are important, but they can be perceived by listeners as a crutch. If you are too tied to note cards or a notepad, the chief message your listeners receive isn't that you're thorough but that you're unprepared. Practice carefully enough to be able to decide what you *don't* need in the way of notes, text, and reminders.

Most effective public speakers gradually winnow their mass of notes and words down to an illustrative outline, which in effect *maps* the presentation. This analogy suggests that the high points (main ideas) are clearly marked along with your planned route (what order you will take them), but the special points of interest of your audience are yet to be determined by your interchange with them during the presentation. Do not lock yourself into a rigid organization by thinking that your notes must tell you the exact wording of each transition, for instance. Such rigidity will preclude the dialogue that can make your map come alive for listeners. Condense. Refine. Trust your sense of interaction. Try not to memorize: If you lose track of where you are in a formal script, you might go completely blank.

POLISHING ORAL PRESENTATIONS

1. Troubleshoot your word choices.
   - Jargon.
   - Needed definitions.

2. Diversify your dry runs.
   - Practice out loud to yourself with notepad nearby.
   - Tape-record and listen.
   - Practice in front of others.

3. Coordinate your visual aids.
   - Readability and accuracy.
   - Use of charts, maps, and tables to increase audience understanding.

4. Practice questions and answers.
   - Learn how to invite questions.
   - Think about what questions might be asked.
   - Answer questions asked: Check your perception of the question, use your answer to clarify an overall point, and verify with the questioner that your response was adequate.

5. Condense and refine notes.
   - Reduce to outline on cards.
   - Key words and definitions.
   - Allow flexibility.
   - Do not try to memorize.

Polishing an oral presentation

Y ou have been asked by a professional service organization to talk to their members about polishing texts and presentations. Your audience will consist of about 100 accounting and finance professionals in middle management. You will have an overhead projector at your disposal along with a computer if you want to use computer-generated graphics or slides. The presentation is to take 15 minutes.

Task: Using this chapter as the core, prepare such a presentation. Then using the section guidelines for oral presentations, refine your presentation. Ask colleagues and friends to be your test audience for this exercise.

Summary

Communication that audiences perceive as professional and credible does not happen magically. Effective communicators know the value of carefully revising their texts and presentations. This final polishing takes time, but the results are well worth the effort. Remember, the way you present your ideas is as important as what you have to say.

How to Approach Practice Tasks

REVISE A MEMORANDUM

Memorandum

To: Frank James
From: John Jones, ABC Accounting
Date: May 29, 200X
Subject: Profit Analysis

   I have looked over your estimates for 200X and have drawn up a projected income statement. (See Exhibit 1.) As you can see, based on your estimates, your pretax income will be $37,250.

   You also wanted to know what multiple of labor cost you would need to charge to increase your profit to $50,000. Again, I have put this information into a table (Exhibit 2). You will need a multiple of labor of about 2.15 in order to make $50,000, holding everything else constant.

   However, Frank, you might want to consider a few issues before you raise your labor charges. Will you lose customers with this higher rate? Will you still be competitive with other repair shops in your area? Are there areas where you can cut costs without changing overall quality? What is the effect of advertising? Perhaps we should discuss these issues before you make any decisions since there might be other ways for you to achieve your desired results without charging more for labor.

   If you have any questions about what I have sent you, please feel free to call me. I will be happy to assist you in any way I can.

Exhibit 1
Projected Income Statement
James Auto Repair

Revenue	Materials	$ 58,200
	Labor	175,950
	Total	$234,150
Expenses	Materials	$ 48,500
	Payroll	103,500
	Rent	7,500
	Utilities	14,300
	Insurance	6,500
	Overhead	16,600
	Total	$196,900
Profit	Revenue	$234,150
	Expenses	−196,900
	Profit	$ 37,250

Exhibit 2
Labor Multiplier
James Auto Repair

If you want to make $50,000, you will need an additional $12,750 in profits. Thus, besides what you now get from labor ($175,950), you need an additional $12,750—a total of $188,700.

Labor cost is $103,500 with only 85 percent billable, or a total of $87,975.

Finally, $188,700 divided by $87,975 yields 2.15, the multiple.

POLISHING AN ORAL PRESENTATION

As you plan this presentation, keep in mind the following. Since your audience will consist of accounting and finance professionals, plan to focus on writing and presentation situations these people are likely to encounter. You may want to develop a scenario similar to ones in this book: Have your audience imagine they have been asked to perform a particular job-related writing or speaking task. Remember your topic is "polishing," so quickly set the scene and then move on to discuss your main message. You can use the information from this chapter and even incorporate material from the text as overhead slides or computer graphics.

You have a time constraint of only 15 minutes, so you will need to plan what you are going to say carefully, making the best use of your time. Follow the suggestions in this chapter to polish and refine your presentation. If possible, do a practice run-through and have a friend critique your performance. You can also use a tape recorder or video camera to help you work out any problem areas in the text. Be sure to pay careful attention to any distracting habits you may have when speaking, and concentrate on eliminating them.

In order for you to be credible to your audience, given this particular topic, you will need to make sure that your presentation is as smooth as you can make it. Of course, this is the goal of any presentation, no matter what the topic.

Endnote

1. R. J. Doolittle, *Professionally Speaking* (Glenview, IL: Scott, Foresman, 1988), p. 118.

Part 3

Focused Communication Contexts

> **66** ... accountants have a responsibility to: communicate information fairly and objectively [and] disclose fully all relevant information that could reasonably be expected to influence an intended user's understanding of the reports, comments, and recommendations presented. **99**

ACCOUNTING PRINCIPLES BOARD (AMERICAN INSTITUTE OF CERTIFIED PUBLIC ACCOUNTANTS, 1970)

Chapter 7

Strategies for Specific Written Communication Formats

"SINCE MUCH OF YOUR TIME WILL BE SPENT WRITING, I'VE BEEN ASSIGNED TO HELP YOU CLEAN UP YOUR PROSE."

Y ou have just gotten to work on Monday morning when the phone rings. It's your cousin Lisa, who is an accounting major at the local university. She is calling to complain about having to write a case study for her managerial accounting course. "I'm really having trouble with this assignment," she complains. "Actually, I think it's just a bunch of busywork anyway. I'm training to be an accountant; how much writing can there be in the real world?" You offer your sympathy, but you also think about your schedule for the week. You have a written report due on Friday, several clients who need to hear from you by letter before the end of the week, and numerous e-mail messages that need to be read and responded to. "Well, Lisa," you say, "based on what I have to do in a normal week, I'd say your assumptions about the lack of writing in the real world are unrealistic. It looks like 80 percent of my time this week will be spent doing some sort of writing."

In the preceding chapters we have discussed general strategies for effective written and oral communication. This chapter and the one to follow focus on specific communication formats that virtually all accountants are expected to master. This chapter reviews the writing process and offers suggestions for communicating effectively within the following written communication formats: letters, memoranda, e-mail, proposals, reports, cases, and group writing projects.

Writing as a Process

Whatever your writing task, your text will be more effective if you remember that effective writing involves four stages: planning, drafting, revising, and proofreading. We have discussed each of these stages in detail in earlier chapters. Before considering specific written communication formats, briefly review the overall process of writing.

1. **Planning:** Develop your communication plan by answering the following questions:
   —**Purpose:** What am I trying to accomplish?
   —**Audience:** Who will be reading this?
   —**Content:** What do I have to say, and how am I going to organize my ideas?
2. **Drafting:** Get your ideas down on paper without worrying about superficial details of correctness and polish. This is only a draft.

THE WRITING PROCESS

- Plan
- Draft
- Revise
- Proofread

3. **Revising:** Look at your text from the reader's point of view. Clarify and reorganize your sentences and paragraphs. Use your communication plan as a guide for revision decisions.

4. **Proofreading:** Correct surface errors in spelling, usage, and punctuation. Also, consider the overall appearance of the document by assessing page layout, spacing, and readability of the typeface. This is the final "quality control" check.

Letters

Business letters are a basic form of writing every accountant will work with on a regular basis. Our suggestions focus on letters written for an audience *outside* of your firm. You might send a letter to your department manager, but this is less common than sending letters out of the office. Of course, you can adapt these

---

BLOCK FORMAT FOR LETTERS

August 24, 200X

Ms. Miranda Stevens
538 Wilson Street
Baltimore, MD 21223

Dear Ms. Stevens:

Thank you for your recent letter inquiring about job opportunities at Rejis Accounting. Your resume is quite impressive; however, at this time we are not anticipating any openings in our accounting staff.

We will keep your resume on file and will contact you if a suitable position becomes available. Again, thank you for your interest in Rejis Accounting.

Sincerely,

Roger Tucker
Personnel Director

[Notice that all lines of the letter align on the left margin in this format. Most sections are separated by a double space with the exception of the closing and the typed name. Here you want to leave 4 spaces to allow for the written signature.]

MODIFIED BLOCK FORMAT FOR LETTERS

August 24, 200X

Ms. Miranda Stevens
538 Wilson Street
Baltimore, MD 21223

Dear Ms. Stevens:

Thank you for your recent letter inquiring about job opportunities at Rejis Accounting. Your resume is quite impressive; however, at this time we are not anticipating any openings in our accounting staff.

We will keep your resume on file and will contact you if a suitable position becomes available. Again, thank you for your interest in Rejis Accounting.

Sincerely,

RogerTucker
Personnel Director

[In this format the date and closing sections are indented and aligned with one another. The first line of each paragraph is indented.]

---

strategies to intraoffice letters simply by taking into account the needs of each specific audience.

Letters should follow a specified, consistent format. Many companies adopt certain styles, sometimes coordinated with the layout of their stationery. Date format and placement, the addressee's name and address, opening salutation, and closing should all be coordinated to enhance the professional appearance of a letter. We have included examples of two letter formats. You can choose either style; neither is inherently preferable. The key is to be consistent within the format you have chosen.

BUSINESS LETTERS: GENERAL ADVICE

Every letter sent reflects on you and on the organization you represent. Some readers will not have met you; they will judge you and our organization largely on the text you create. Have you ever received a business letter from someone you did not know? By the time you finished reading the letter, you probably formed a positive or negative mental image of the author and his or her organization. The same is true for the letters you send. Therefore, all business letters should encourage the reader to have a positive attitude toward doing business with you and your company. Remember, too, that readers will react not only to *what* you say but to *how* you say it. A letter that sounds cold and impersonal or a letter that is disorganized, poorly written, or full of grammatical mistakes can have a negative impact on your relationship with your reader. Many professionals tell stories about firms with whom they will not do business because of poorly written letters.

The letters you write as an accountant will have many possible goals, including providing information (answering tax questions for a client), requesting information (asking a software manufacturer to explain a function in their spreadsheet program), requesting action (asking a client to choose from among suggested alternatives), and responding to requests (agreeing with the IRS to an in-house audit for a client).

Although your reasons for writing a business letter will vary, seven general strategies can guide you in all situations. Obviously, however, you must temper the following advice by considering other factors, including audience analysis.

1. **Be explicit.** Make your point quickly; do not make the reader struggle through a long, complicated introduction before discovering why you are writing.
   Inexact: I am writing to you to discuss the fiscal plan that you recently sent to me to evaluate for you. My results can be summarized on the income statement that is attached. The results that I came up with, based on the information you gave me, indicate that your company will end up with a pretax profit of $37,250.
   Explicit: I analyzed your fiscal plan; based on your data, your pretax profit will be $37,250. See the attached pro forma income statement for details and assumptions.

2. **Be concise.** Focus on what you have to say, and delete sections that are wordy or not pertinent.
   Wordy: As per your request, I wish to inform you that an analysis of information indicates that the tax regulations in France do not appear to produce the same benefits as in Japan.
   Concise: In conducting the analysis you requested, I found tax regulations are more beneficial in Japan than in France.

3. **Be precise and specific.** Avoid sentences that may be ambiguous or too abstract.
   Ambiguous: We have had trouble getting a response from Ajax Company or Frank, Inc. They do not communicate with us or Jones. This makes it a problem to proceed with the project.
   Specific: Because neither Ajax Company nor Frank, Inc. has responded to our letters or those from Jones, we cannot proceed with the project.

4. **Maintain a friendly, professional tone.** Avoid sounding brusque or cold. One simple way to establish a personalized tone is to write in the first person, directly addressing the reader. Depending on your relationship with,

for example, Helen Corazon, such references may be "Ms. Corazon" or "Helen." Using "please" and "thank you" can also help to convey a friendly, polite tone.

Unfriendly: Our office is in receipt of your correspondence. Be advised that several measures exist that can be instituted to bring about a favorable resolution to your fiscal crisis.

Personal: I received your letter, John, and I have some ideas to help you get through this fiscal crisis.

Another way to create a more personal tone is to use the first person singular pronoun "I" when referring to yourself or the first person plural pronoun "we" when referring to your company.

Impersonal: Your recent letter has been received by this office. The writer will be handling your audit, so direct any further correspondence to her.

Personal: We received your letter and appreciate your concern. I will be conducting your audit, so please feel free to call me if you have any questions or concerns.

You can choose the singular or plural first person pronoun based on the specific context of the writing situation.

5. **Use a conversational style.** Create short sentences and paragraphs, using the natural language of conversation. Read your letters out loud. Do your words and sentences sound natural? Are you using phrases that resemble normal speaking patterns? If your letters sound stilted and unnatural to you, they will likely sound that way to your readers.

   Unnatural: Pursuant to our recent conversation by telephone, enclosed please find an analysis developed to ascertain the feasibility of the utilization of straight-line depreciation.

   Natural: As we discussed on the phone, I am enclosing an analysis of whether you should use straight-line depreciation.

6. **Close on a positive, friendly note.** Your ending should motivate the reader to do business with your firm in the future. Depending on the situation, you might want to thank the person for considering your ideas or indicate your availability for continued discussion. No matter what the context, your closing should be the equivalent of a polite "good-bye."

   Poor closing: Please be advised that despite the fact that our proposal is delayed, someone from our office will telephone you next week with the complete plan.

   Positive closing: Thank you for your patience. I apologize for the delay in sending you our proposal and will call you Monday with the complete plan.

7. **Use an appropriate level of formality.** Audience analysis should help you decide how formal the letter should be. Formality is achieved not only by elevated language and syntax choices but also by avoiding contractions and

---

GENERAL STRATEGIES FOR BUSINESS LETTERS

- Be explicit.
- Be concise.
- Be precise and specific.
- Maintain a friendly, professional tone.
- Use a conversational style.
- Close on a positive, friendly note.
- Use an appropriate level of formality.

---

by addressing the reader as Mr. or Ms. in the greeting. Usually the first time you write to a person you do not know, you will want to choose this type of formal style. As further communication develops with this person, you can loosen up your style by using a first name and switching to contractions and other less formal word choices.

Formal: Dear Mr. Carlson: To date I have not been able to arrange a time for our two groups to meet. This week will not work. What is your schedule for next week?

Informal: Dear Jim: I haven't been able to set up a meeting time. I know this week won't work for our group. How about next week?

BUSINESS LETTERS: SPECIFIC CONTEXTS

The following are specific strategies for writing effectively within the various contexts you may encounter in business communication.

1. **"Good news" letters** (letters welcomed by the reader—awarding contracts, reporting on a favorable ruling by the IRS, etc.)
    —Put the good news or main idea first. ("I have just received the IRS report, and you will be pleased to know they have decided not to audit the New York office.")
    —Follow with the necessary details or explanations. ("The IRS decision was based on the following factors: . . . ")
    —End on a positive note by:
      –Recalling benefits of the good news. ("Avoiding this audit means we can turn our full attention to the year-end closing.")
      –Expressing appreciation. ("Thank you for taking the extra time to prepare for this review.")
      –Motivating action. ("Now that this decision has been made, let's begin planning for next year's diversification.")
      –Expressing willingness to help further. ("I will be happy to discuss future plans at your convenience.")

2. **"Bad news" letters** (negative message for the reader—denial of a request, complaint, reprimand, objections, disagreements, etc.)

 A. Indirect approach (delaying the "bad news")
   - Prepare the reader for bad news by using the first sentence or paragraph as a "buffer" by:
     —Providing information to help the reader accept the message in its proper context. ("Trends in business seem to follow cycles; some downturns are predictable while others seem to develop for no apparent reason. We should remember the long-term picture when assessing quarterly reports. Unfortunately, . . . ")
     —Expressing understanding of the reader's need or problem. ("I can understand your desire to lower your break-even point in new product development. However, . . . ")
     —Showing you have taken the reader's situation seriously. ("I have analyzed these figures carefully, using a variety of approaches. Despite this . . . ")
     —Expressing appreciation for the reader's interest. ("Thank you for your recent request. Unfortunately, . . . ")
   - Explain the reasons for the unfavorable news, then state the bad news. ("The committee awarded this contract to the group that most closely adhered to the prescribed format. Based on this . . . ")
   - End the letter on a positive note by:
     —Expressing appreciation to the reader. ("Thank you for your thoughtful response.")
     —Expressing willingness to cooperate in the future. ("Although the budget for next year is going to be difficult to achieve, I will monitor each division on a quarterly basis to keep you up to date.")
     —Expressing continued interest in the reader. ("Although your bid was not accepted on this project, we will be pleased to receive your bids on future projects.")

 B. Direct approach
   - Announce the bad news diplomatically in the first sentence or first paragraph. ("After careful consideration, we have decided not to review new general ledger software during the upcoming fiscal year.")
   - Make the reasons for the decision clear. ("This decision resulted from the following factors: . . . ")
   - End on a positive note by:
     —Expressing appreciation for the reader. ("Again, thank you for your time and cooperation.")
     —Expressing willingness to cooperate in the future. ("Please remember we will contact you when we are ready to proceed with this project next year.")
     —Expressing continued interest in the reader. ("Although we have decided against looking at new software this year, we will certainly consider your firm if we change our stance on this issue.")

3. **Letters of request or inquiry** (seeking information, proposals, permission, etc.)
    —State your request in the first sentence or first paragraph. ("In looking through our records, we are missing the monthly statement for February. Please send a copy of this report as soon as possible.")
    —Help your reader think through your request by discussing the implications of your request. ("This monthly statement is crucial to the budget planning process already under way.")
    —Close with a simple thank you. ("Thank you for your consideration.")

4. **Letters of persuasion** (sales)
    —Establish rapport by gaining the reader's interest and giving your reader a reason to read on. ("Every business is looking to increase profits. Our firm has an approach that has benefited more than 50 companies in your area and will work for you.")
    —Begin to develop your selling point. ("Our approach is simple, but effective; small business owners deserve the same quality service as giant corporations.")
    —Consider these opening strategies:
      –Begin with a surprising fact. ("Today, small businesses represent the major hope for sustaining the economy.")
      –Make an emotional appeal. ("The threat of bankruptcy looms large in today's world. Through careful fiscal planning, you can avoid falling victim to bankruptcy.")
      –Create a scenario. ("Imagine you have been asked to develop a new fiscal policy for your organization. Where can you look for help?")
      –Begin with a story or anecdote. ("In 1989, a small, little-known graphics company took a chance on an innovative approach to fiscal planning. Today, this company is one of the leaders in the field of graphic arts.")
      –Offer testimonials. ("When asked to rate the service of our firm, our clients continually give us high marks for caring and concern.")
      –Suggest a problem you will solve. ("If you are like most business owners, you worry about what an IRS audit will find. We can help take that worry away from you.")
      –Identify with the reader. ("We know how it feels to watch overhead costs eat away at your profits.")
      –Compliment the reader. ("Your diligence and hard work have earned you a solid reputation in the small business community. Now you are ready to take the next step.")
      –Ask a question. ("Have you ever considered the advantages of working with an outside accounting firm?")
      –Use a command. ("Do not let another day go by without considering the advantages of choosing our firm to handle all your accounting needs.")
      –Show how to save money. ("Our approach will save you more than twice our fees.")

Indirect or Direct Approach

Sometimes the James Company pays its bills on time, but often payments are quite late. Currently, James owes Shasta Crafts $25,000 in receivables that are 90 days past due. In your role as a member of the accounting staff at Shasta, you have been asked to write James a letter letting them know if the bills are still unpaid after 10 days, Shasta will no longer do business with James and will pursue legal action against them.

Tasks: Write two versions of a letter to Harry James, President, using the indirect method and then the direct method.

—Develop your central selling point and present your other appeals. ("Our method works. We are so sure of our approach that we guarantee you will see a difference by the end of the next quarter.")

—Close by:

–Asking the reader to act within a limited time. ("The sooner you decide, the sooner you will see the benefits. Call today.")

–Emphasizing reader benefits. ("Remember, your company will profit from using our services.")

MEMORANDA AND ELECTRONIC MAIL

Business communicators commonly use memos and e-mail as messages written for readers *within* an organization. By nature, memoranda and electronic mail are less formal than letters, may presume multiple readership, and usually assume your readers know each other. Even so, the design of a memo or an e-mail message is no less important than that of a letter. Memos and e-mail are the major ways that professionals communicate in writing with peers, supervisors, and subordinates.

General advice. Memos and e-mail provide information about operations and influence decisions. Information in memos can flow *up, across,* and *down* in an organization; memos and e-mail make you visible within your organization. Your memos and e-mail may have a wide range of goals, such as making announcements, giving instructions, confirming agreements, clarifying information, giving reminders, promoting goodwill, reporting bad news, and persuading.

Readers generally have two attitudes toward memos and e-mail: They are receptive to messages that give necessary information, or request information they are able and willing to give; or they are resistant to messages that give bad news or make recommendations they are reluctant to follow.

Some general guidelines include:

1. Assess your reader(s) based on the following:
   —Your business relationship with the reader.
   —Your reader's knowledge of the subject.
   —Your reader's attitude toward you and toward the subject of the memo.

GENERAL STRATEGIES FOR MEMOS AND ELECTRONIC MAIL

- Assess your readers.
- Create a header that clearly announces audience and purpose.
- Maintain a cordial and respectful tone.
- Use a conversational style, showing courtesy and sympathy.
- Proofread your text carefully.

---

2. Create a header that clearly announces the audience and purpose for the memo. For example:

TO:	Budget Committee
FROM:	Regina Allwell
SUBJECT:	Friday Deadline
DATE:	February 23, 200X

   It is appropriate to put your title if the intended audience is not familiar with you or if organizational style calls for it.

3. Maintain a cordial and respectful tone.
   —Remove negative expressions and statements that may anger or alienate your reader.
   —Choose language and information that stresses the benefits of your message.

4. Use a conversational style, showing courtesy and sympathy.
   —Strive for a balance of "I" and "you."
   —Use short paragraphs and sentences.
   —Use words and phrases you would use in conversation.
   —Use contractions.

5. Proofread your text carefully.
   —Use the proofreading methods discussed in Chapter 6.
   —Remember, your readers will form an opinion about your professionalism by the polish of the texts you send them. You want your messages to project a positive image.

Memoranda: specific contexts

1. Memos that give or request information
   —State the purpose of the memo. ("Our department needs the final figures from the budget committee by the end of the week.")
   —Present relevant facts or examples. ("We cannot proceed without this data, and the home office is pressuring us for a response.")
   —Close with a cordial remark; ask for action. ("Thanks for your help; remember, the deadline is Friday.")

2. "Bad news" or persuasive memos
   —Establish a shared goal or common frame of reference. ("We all know how difficult the last few months have been.")

Writing Memos

Write a memorandum telling the recipient you will be gone from the office next week and have to cancel an important meeting between the two of you. Use the following roles and audiences:

Tasks:

1. Middle-level manager writing to a corporate vice president.

2. Corporate vice president writing to a middle-level manager.

—Provide information that will support the bad news. ("Despite our careful planning, the new product is not selling well. We need to make additional changes.")

—Give the bad news. ("We have decided to delay the hiring of new staff members.")

—Close with a cordial remark; ask for or promise action. ("Although the hiring is being delayed, I hope each of you will do all you can to help us get through the next few months.")

3. **Memos that promote goodwill**

—Begin by announcing your message. ("Congratulations to Mavis Burke, our new senior partner.")

—Give some significant details that explain why you are offering praise, thanks, or sympathy. ("Thanks to your efforts we have signed 10 new clients this month.")

—End by reinforcing your positive attitude toward the reader. ("We couldn't have done it without everyone's hard work. Thanks to you all.")

Electronic mail: specific considerations. Today's rapid-fire business environment requires all accountants to become comfortable with a writing context that was unnecessary or peripheral only a few years ago. The advent of electronic mail has changed some basic assumptions about writing. The e-mail message is not simply a memorandum transmitted electronically. This new medium brings with it different assumptions.

In some ways, those assumptions make e-mail communication more like everyday speech than most corporate writing practices. E-mail etiquette demands a less formal, more immediate, more vivid, and more direct writing style than previous contexts presume. It also has its own special stylistic conventions that you will want to understand.

Some basic differences between e-mail and other writing forms:

• E-mail is simultaneously less formal and more terse than memos, and most communicators expect succinct messages that are nevertheless friendly. E-mail messages put a premium on relaying simple information like a minor change in office policy, or reaching agreements quickly on such matters as coordinating where to hold a meeting with colleagues, or negotiating a deadline for a report with a busy superior. Many e-mails, therefore, take advantage of bullet point lists of sentence fragments, or numbered lists embedded in short paragraphs, when communicating information, and simple, clear inquiries when requesting clarification. E-mail is not a particularly good medium for extended analysis of complex problems. Save that task for memoranda.

- E-mail in organizations is potentially more public than most memos, and most communicators are aware that private or secret sentiments can be monitored by superiors. Be careful what you write. Most systems save even deleted e-mail messages.

Some important conventions and idiosyncrasies of e-mail writing include[1]:

1. **Keep your message short.** For most messages, the reader should be able to see the entire content (including your header) on one screen.

2. **Compose a subject line carefully.** The *subject line* is where you announce the topic or reason motivating your message. Because many executives receive dozens of e-mails daily, they often prioritize and read them according to how important it is to read and respond right away. An e-mail with a subject line that says "Some ideas" is likely to receive less immediate attention than one that announces "Friday's deadline for new purchase proposals." Make sure to tell your story in the subject line.

3. **Use "cc" (carbon copy or complimentary copy) judiciously.** The *cc* e-mail command automatically sends a copy of the message to designated other receivers in addition to your main correspondent. Some are necessary recipients, but do not overdo this option. Most of your colleagues are already drowning in information clutter. Send mail only to those who have an active need for the information.

4. **Write with natural, conversational language whenever possible.** This is not e-mail language: "I note that we, as a vigorous corporation committed to public service, have not done much to invigorate this community during the past decade. I would charge your group with investigating the potential for further investments in community-based resources." This is e-mail language: "We've done a lot for the community, but should do much more. Jane, could you ask your group for ideas?"

5. **Know basic etiquette of e-mail.**
   —TYPING IN ALL CAPS is usually considered the e-mail equivalent of shouting and often frowned upon.
   —*Emoticons,* also known as *smileys,* are some people's way of expressing nonverbal-like emotional messages to accompany the words (so named because they can look like smiles, winks, frowns, and the like, viewed sideways [:-) ;-) {:-( ]). Overenthusiastic use of smileys can grate on experienced e-mailers, so use them with care.

SOME E-MAIL CONSIDERATIONS

Some basic differences between e-mail and other writing forms:

- E-mail is simultaneously less formal and more terse than memos.
- E-mail is potentially more public than most memos.

Some important conventions and idiosyncrasies:

- Keep your message short.
- Compose a subject line carefully.
- Use "cc" (carbon copy or complimentary copy) judiciously.
- Write with speechlike language whenever possible.
- Know basic etiquette.
- Do not be in a hurry to respond.
- Consider carefully what parts of an original message to include when responding to it.
- Know what software your e-mail recipients have for files you attach.

—Be especially careful also of the tone of negative messages or bad news; on-line jargon has a special word for posts that are too dismissive or insulting: *flaming.*

—Although all the basic information identifying yourself is included in the header of e-mail, most recipients appreciate both a brief greeting and a brief good-bye message to personalize things.

—Be careful not to be too "cute" in messages. This applies to the use of things like colored type or clip art.

6. **Sometimes it's better to wait before responding.** If you are dealing with a subject that might be touchy to a person who has just sent you an e-mail or touchy to you, wait before responding. It's tempting to just shoot back a quick answer. As with letters, it's good to draft a response and consider it again the next day before sending it. Also, there are times when a response may not be needed, such as a message sent to an entire department.

7. **When replying to a prior message, include quoted material to which you refer, but only the specific parts to which you are replying.** Do not reproduce all of the other person's lengthy e-mail if you only want to add a reservation or warning to the third paragraph, for example. Having to reread long sections from the original message wastes your correspondent's time. Use your software editing capability to include only relevant quotes from the prior e-mail, and streamline your reply while making clear your response.

8. **Know what software recipients have for files you attach.** Remember that although your e-mail software may allow you to attach and send files with your basic posts, not everyone will have the right software to open those files once they are downloaded.

Proposals

Proposals are persuasive reports written to sell a product, service, or idea to a given audience. Effective proposals must convince readers that you can provide a product, service, or idea that will meet their specific needs. Since your proposals will often compete with those from other organizations or those from other staff members, a strong proposal should demonstrate the advantages of your offering relative to other alternatives.

PROPOSALS: GENERAL ADVICE

1. **Analyze the needs of your reader(s).** Before you can convince your readers to accept your ideas or your services, you need to understand their specific needs and concerns.

2. **Know your information.** Readers are persuaded when you sound knowledgeable and capable. Do your homework by researching your topic thoroughly. Use specific examples and direct language to assure your readers of your expertise. Provide ideas your readers do not already know, or place common ideas in novel contexts.

3. **Choose a logical format.** Executives tend to be persuaded by arguments that build logically. Most proposals follow a three-step format:
   —State the problem.
   —Analyze the situation.
   —Present the solution and recommendations.

4. **Use headings to aid readability.** Headings help readers see the overall organization of your material. They also help you as a writer to focus and structure your material. In longer proposals, a brief Table of Contents (usually an outline of your first-level and second-level headings) is helpful. Below are some typical headings used in proposals:

I. Introduction
    Subject and Purpose
    Background
    Problem Statement
II. Proposed Plan
    Methods
    Feasibility
    Budget
III. Conclusion

      The headings listed above are general, and you will need to use headings appropriate for your specific proposal. For example, you may want to use the more specific heading "What Went Wrong at Nigel Manufacturing" instead of the more generic heading "Problem Statement."

5. **Anticipate questions and objections.** Think about possible concerns or negative reactions your readers may have toward your proposal. You might want to acknowledge these concerns in the text and then provide solutions to overcome their objections. Perhaps when you were interviewing the client, you got the impression that one of the key decision makers had specific reasons why she was cool to the basic idea under discussion. This is an opportunity; use it to develop your proposal in an assertive (not defensive) manner. If you know you are in competition with another group submitting a proposal, you can discuss the unique advantages your organization offers.

6. **Provide solid evidence to support your recommendations.** Make certain your solutions are built on a logical approach and are based on current data. Business leaders respect data that are empirical (related to observable events), concrete, and specific. If your recommendations are based on a certain adverse trend, do not merely *claim* there is a trend. Supply the data to *demonstrate* the trend.

7. **Provide recommendations that address all aspects of the problem statement.** If you have listed three areas of concern in your opening problem statement, you must provide recommendations to solve all three problems. Your "problem statement" and "recommendation" sections should mirror one another.

8. **Use graphics to aid comprehension and to enhance the text.** Since graphics provide a visual explanation of a process or trend, you may want to incorporate graphic elements in your proposals. Keep your graphic elements close to the text discussing them. Most word processing programs allow you to insert graphs directly into the text. Choose a style and size that fit the impact you want. If close placement is not possible, be sure to tell the reader exactly where the visual is located. For example: (See Exhibit A on page 5.).

9. **Use appendixes for information that would interrupt the text.** Long, complicated tables and charts should be placed at the end of the proposal in order to keep the flow of your text intact. Again, be sure to tell the reader exactly where the information has been placed. This goes with the thought that tables are useful when dealing with details. In almost all forms of communication, you do not want to include details like the following:

   First we took sales and multiplied by an inflation factor of 1.03. Then we deducted an estimate of bad debts (2 percent). We then multiplied this by . . .

   Details such as these, if appropriate to include at all, should be part of an exhibit.

10. **Proofread your final copy with extra care.** Recently, a company asked several engineering firms to submit proposals on a major consulting job. One of the eight proposals contained several typographical and spelling errors. Those evaluating the proposals not only rejected this engineering firm solely on the basis of the errors in the proposal but also decided they would not ask this firm to bid on future work. Thus, errors can have a long-lasting effect.

> ## GENERAL STRATEGIES FOR PROPOSALS
>
> • Analyze the needs of your readers.
> • Know your information.
> • Choose a logical format.
> • Use headings to aid readability.
> • Anticipate questions and objections.
> • Provide solid evidence to support your recommendations.
> • Provide recommendations that address all aspects of the problem statement.
> • Use graphics to aid comprehension and to enhance the text.
> • Use appendixes for information that would interrupt the text.
> • Proofread your final copy with extra care.

Reports

Reports are written for both internal and external audiences. They might be sent with a cover letter or memo, or they might be hand-delivered as part of a presentation.

The reports you write as an accountant generally fall into two categories: *information reports* and *recommendation reports.* You might be asked to provide alternative ways to report an event for financial reporting (information) and/or to recommend a particular course of action given the alternatives. Many of the suggestions listed for proposals also apply to the writing of reports.

GENERAL ADVICE

1. **Know your audience.** Think about your readers' interest in this report and what their attitudes are likely to be. As with every audience analysis, adapt your information to the level of expertise of your readers. In addition, plan ahead to answer any possible objections your readers may have to the information you are presenting.

2. **Use headings.** Use both major and minor headings when appropriate. Tailor headings to fit the content and context of the report; your objective is to guide the reader through the flow of your presentation. Consider one typical way to organize a report:

   I. Title Page

  II. Table of Contents

 III. Executive Summary

 IV. Introduction

  V. Methodology

 VI. Results

VII. Limitations

VIII. Conclusions and Recommendations

 IX. Appendixes

Title page. A title page includes (1) the title of the report (choose a title that reflects the precise nature of the information), (2) the group for whom the report was prepared, (3) the author(s), and (4) the date of release or presentation. If the report is for limited distribution, this fact should be noted on the title page; a list of people to whom the report should be circulated may also be supplied. Many companies have preprinted title pages with specific places for basic information.

Table of contents. The table of contents includes (1) the titles and page numbers of major sections and subtopics within each section and (2) a list of all appendixes. If a number of charts and tables are used, they should be listed on a separate page immediately following the table of contents.

Executive summary. Because an executive summary includes a brief description of (1) objectives, (2) results, and (3) conclusions and recommendations, it is the most important part of the report.

Introduction. The introduction to a report can include (1) background material, (2) a clear statement of objectives, and (3) an overview of the organization of the report. Introductions set the stage for all that follows. You want to provide appropriate background information and to let the readers know what you are going to do, how you are going to do it, and what you want them to learn from the report. The first section of the introduction should include a description of the problem and the factors that influence it. The objectives originate from the background information, but they need to be stated explicitly. You might even give a brief statement of your conclusions.

Methodology. In some sense, almost all reports involve research—for example, a particular tax treatment, alternate software to purchase, a business plan. As such, you will have used a specific methodology to reach your conclusions. Thus, in this section include a summary of the methods used to meet the objectives of the project. Keep technical details to a minimum. Where necessary, place such details in appendixes.

Results. This section includes a meaningful description of research findings, organized around the objectives of the study. This section should not be a long series of statistical tables. Use carefully chosen summary tables and visual aids to clarify the discussion.

Limitations. This section should highlight (1) problems in the research and (2) potential limitations of the data. Point out any shortcomings that your readers might overlook, such as the danger of generalizing a local study to the national market.

Conclusions and recommendations. This final section can include conclusions drawn from the research and stated in relation to each objective of the project. State each objective and then present the specific conclusions relevant to that objective. Depending on the specific context, recommendations can be made if appropriate. This section is also the appropriate place to suggest additional research that might be undertaken.

Appendixes. Use appendixes for items, such as photocopies of relevant memos or newspaper articles, that will appeal to only a few readers or that may be needed only for occasional reference.

EXECUTIVE SUMMARIES

An executive summary is the skeleton of a report, presenting only the *essential* information. It provides a basic sense of the entire report and allows the reader to decide if the entire document needs to be read. For many of your readers, the executive summary is the *only* part of a report they may review.

Depending on the report topic, an executive summary may contain all or some of the following elements:

1. **Purpose and scope of the report**—why the report was written and what the report covers.
2. **Methodology**—description of research used.
3. **Findings**—what was learned.
4. **Conclusions and recommendations**—what can be inferred from the research and what should be done on the basis of these conclusions.
5. **Implementation**—feasibility and steps needed to implement the recommendations.

In some contexts, the executive summary is similar to a table of contents because it allows a reader to refer directly to topics that bear closer reading. For example, lawyers often provide their clients with summaries of long documents. Not only can the summary be used to review salient points, but also the client can identify potential problem areas for detailed discussion.

You have a choice of formats for an executive summary. Some writers prefer to use an outline form (bullet list or bullet points). This format is brief and does not require that points be in full sentences. You literally highlight key phrases in an outline form. Another approach is to use full text and short paragraphs for your summary. Different individuals have different styles, and certain organizations have a preferred style.

EXECUTIVE SUMMARIES: GENERAL ADVICE

1. Remember that an executive summary is *not* the introduction to the report. It is a "capsulized" version of the entire document.
2. You can write a draft of the executive summary either before or after you write the report.
3. If you write a draft *before* you write the report, you will be able to judge if you have a sense of the entire report before you begin putting it together.
4. If you write a draft *after* you write the report, use the report's headings and subheadings to help you identify the major elements.

> ### GENERAL STRATEGIES FOR REPORTS
>
> - Know your audience.
> - Use headings.
> - Consider writing an executive summary.

5. Remember to summarize without using long passages from the text of the report. The idea is to give the reader an *overview* of the document.

6. Keep your summary to one or two pages in length.

7. Ensure that your summary and report are consistent with each other.

Case Study Analyses

Cases emphasize real-world issues and problems in an accounting course. They can be broad and include general strategic issues, or they can be more focused and deal with a specific problem. When you write a case analysis, you are expected to draw on ideas introduced in your textbook and/or other readings as well as information from class lectures. Cases help you to develop both analytical thinking skills and writing skills.

When asked to write a case analysis for a class, you might not be given a specific role but rather will write as a student to the instructor. Obviously, this does not mirror the kinds of roles and audiences you will have as a professional accountant. However, the practice of preparing a case analysis can help you to develop writing and organization skills that you will use as a practicing professional.

Depending on the assignment and directions from your instructor, a case might be cast as a simulation of a real-world report or extended memorandum with you playing the role of an outside consultant or a member of a company's staff writing to specific managers within the firm. In this context, refer to our suggestions about letters and memos (you might want to attach a cover letter) and reports (the case analysis itself). The instructor might want a specific format (e.g., requiring an executive summary).

In a traditional academic analysis, clarify the basic form, format, and coverage requirements with your instructor. For example, you will need to know if you are to create a formal, third-person analysis or if a less formal approach is acceptable. Clarifying these requirements is no different than seeking information from a supervisor in a work setting.

GENERAL ADVICE

1. Read the case thoroughly. Read carefully to get an overall understanding of the situation and the supporting facts. Then, reread the case, taking notes about key facts, assumptions, and issues. If case questions are provided, use them to formulate an approach to your analysis.

> ### GENERAL STRATEGIES FOR CASE STUDY ANALYSES
> ....
> - Read the case thoroughly.
> - Develop a plan or format with appropriate headings.
> - Analyze rather than describe.

2. **Develop a plan or format with appropriate headings.** Decide on an organizational scheme and use headings to delineate your sections. A traditional case analysis would include:

   I. Executive Summary

   II. Introduction

   III. Internal Analysis (factors within the organization)

   IV. External Analysis (factors outside the organization)

   V. Conclusions/Recommendations

   VI. Summary

3. **Analyze rather than describe.** A case assignment requires that you go beyond restating the facts of the case (description). You will be evaluated on how well you use the information and analyze its implications. This requirement applies no matter what format you choose. See Chapter 4 for a detailed discussion of analysis.

Group Writing Projects

At some time during your career you may find yourself participating in a group writing project. Many companies are currently stressing teamwork and cooperation, and group projects are a natural result of this emphasis. Group authorship has the advantage of dividing the work among several people rather than having one person be responsible for the entire text. Group writing projects also allow the various points of view and areas of expertise of the group members to be incorporated in the final product. Of course there are also dangers in group writing. A successful project depends on careful planning and the complete cooperation of all group members. Individuals who do not follow the agreed-upon approach or who do not meet deadlines can endanger the entire project. In addition the final text needs to sound as though it were written by one author rather than by a group. This requires careful editing for style and consistency. With careful planning and supervision, being part of a group writing project can be an interesting and rewarding experience.

GENERAL ADVICE

1. **Set up the project.** Call an initial meeting to chart your course. Determine and agree on purpose. Analyze the audience. Generate ideas. Develop a working outline. Assign section(s) to each group member to research and write, and make sure all persons fully understand their roles.

2. **Discuss individual expectations.** Each group member should be honest about the amount of time available for the project. If you know you will be able to devote only two or three hours per week, then you need to say that and not take on duties that will require many extra hours. Members should regard their early agreements as commitments, to be changed later only with the full knowledge of the rest of the group.

3. **Develop a realistic schedule for completion of tasks.** Establish a timetable that is reasonable but will meet project deadlines. Project success depends on sticking to individual deadlines; every group member needs to understand the importance of having work done on time. It's usually helpful to build into the schedule a day or two of contingency time to allow the group to respond to last-minute emergencies.

4. **Adjust the outline, tasks, expectations, and schedule as appropriate.** Conditions might change. Maintain contact with each other so that you can adjust plans as new ideas occur and as deadlines are modified. For example, if a team member should get sick or have a personal emergency, others would then need to readjust the project, tasks, and schedule accordingly.

5. **Revise the text to achieve consistency and a single voice.** You can accomplish this essential but difficult element of group writing by choosing one group member or someone outside the group to serve as overall editor. This person will revise the sections to be consistent in tone, style, and format. The editor's goal is to create a text that flows smoothly from beginning to end.

6. **Avoid common problems in group writing.**
   Problem: Poor project management.
   Solution: Start early, plan thoroughly, and assess progress often.
   Problem: Failure to develop clear purpose or sense of audience.
   Solution: Discuss and agree on purpose and audience before beginning.
   Problem: Failure to achieve a single "voice."
   Solution: Appoint one person to coordinate stylistic consistency and have other group members serve as supplemental editors.

One of the challenges of working together involves the group's power structure. In some cases, there will be an official team leader. For example, in public accounting a manager or senior staff person will be placed in charge. When all team members are peers, then the team develops its own leader(s). Communication among members is crucial so that problems are addressed promptly. For example, in some teams an individual or a clique takes over the team. Sometimes one or two people want to do virtually all the work, thereby limiting others' participation and overlooking the creative ideas they might provide. Try to work out problems within the group. However, if you cannot resolve things, members should consult with whoever is supervising the group, rather than letting the situation get out of control.

> ### GENERAL STRATEGIES FOR GROUP WRITING PROJECTS
>
> * Set up the project.
> * Discuss individual expectations.
> * Develop a realistic schedule for completion of tasks.
> * Adjust the outline, tasks, expectations, and schedule as appropriate.
> * Revise the text to achieve consistency and a single voice.
> * Avoid common problems in group writing.

Summary

All accountants must communicate their expertise through writing; composing letters, memos, and reports often dominates an accountant's workweek. Your career success and advancement depend in large part on your writing abilities. Remember, writing is a skill that improves and comes more naturally with practice. The best way to strengthen your written communication skills is to write, and write often.

How to Approach Practice Tasks

INDIRECT OR DIRECT APPROACH

Indirect method:

> May 14, 200X
>
> Mr. Harry James, President
> James Company
> 156 East Main Street
> New Concord, Ohio 43762
>
> Dear Mr. James:
>
> We at Shasta Crafts are proud of our service and delivery records. We appreciate being chosen to supply your needs. Unfortunately, James Company presently has invoice amounts totaling $25,000 that are 90 days past due.
>
> We value you as a customer, but your record of payment has been sporadic in the past. Please send us the outstanding amount by May 24. If we do not receive payment by this date, we will have no other choice but to seek legal action.
>
> Thank you for taking care of this situation. Please give me a call if you have any questions.
>
> Sincerely,

Direct method:

May 14, 200X

Mr. Harry James, President
James Company
156 East Main Street
New Concord, Ohio 43762

Dear Mr. James:

James Company has outstanding invoice amounts totaling $25,000. These bills are presently 90 days past due.

Please send the outstanding amount by May 24. If we do not receive payment on this date, we will terminate our business relationship with you and will seek legal action to receive payment.

Thank you for addressing this situation.

Sincerely,

WRITING MEMOS

Memorandum

To: Karen Regis, Vice President
From: Bill Gettys, Finance Manager
Subject: Change in Schedule
Date: September 22, 200X

We tentatively scheduled a meeting for next Tuesday, September 27, to discuss proposed software purchases for the finance department. Unfortunately, I am no longer free on that date. The Dallas office has a crisis with their personnel that requires my assistance. I need to fly there on Monday and do not expect to get back before Friday.

I am sorry to postpone our meeting; I know how important these purchasing decisions are. I would like to reschedule at your convenience. Give me a call, so we can set up a new date.

Memorandum

To:	Bill Gettys, Finance Manager
From:	Karen Regis, Vice President
Subject:	Change in Schedule
Date:	September 22, 200X

I'm afraid I need to reschedule the meeting for next Tuesday. I have to be gone from the office all of next week. I know you're anxious to get moving on the software purchases, so let's try and get together on Monday, October 3.

Check your calendar and get back to me as soon as possible. I'm sorry for the inconvenience. I'm looking forward to hearing your ideas.

Endnote

1. There are many sources for hints on e-mail. Some of this material comes from B. Howard, "Avoiding Clueless E-Mail," *PC Magazine,* May 26, 1998, p. 97. Another good source is D. Angell and B. Heslop, *The Elements of E-Mail Style: Communicate Effectively via Electronic Mail* (Reading, MA: Addison-Wesley, 1994).

Chapter 8

Contexts for Oral Communication

F rancine Marti and a team of two other accountants have spent over a month researching the economic implications of closing the Dayton plant. The group met four times last week to discuss their recommendations and how they should be written up. They decided on the report's organization, format, and even the type of binding most appropriate for delivering the manuscript to upper management.

Marti spent over 20 hours at work and at home drafting the text, and another six or seven hours revising it after her colleagues' close reading and careful feedback. The report, finally ready, is on the boss's desk Wednesday, two days ahead of its deadline. It is an imposing achievement: 82 pages of data, interpretations, charts, and recommendations supplemented by several appendixes of policies and precedents. The team of accountants is justifiably proud of it.

Late Wednesday afternoon, each of the three accountants receives a memo requesting a 20-minute "presentation of findings" to company executives on Monday morning in the conference room. The team cannot decide whether this is because the executives are lazy and do not want to read the entire report, whether they are severely pressed for time and want to know firsthand what the team thinks is the nitty-gritty material in the document, or whether they need this oral summary in order to focus their reading more effectively. A project that has consumed five weeks of the three accountants' time is now about to consume five more days. Its ultimate success will depend upon a mere 20 minutes. Therefore, the upcoming five days in the life of the project are just as crucial as the preceding five weeks.

Today's executives are relying more and more on oral presentations. One business executive estimates that 33 million corporate presentations are given each *day*. Xerox estimates that *70 percent* of a typical executive's time is spent in meetings.[1] This does not leave much time for reading 80-page reports carefully. No wonder executives would want to (*a*) hear only those facts and recommendations they need to hear, (*b*) streamline meetings they need to have, and, therefore, (*c*) encourage ever more concise presentations by others.

In other words, a full written report is for the files and perhaps for reference; *an oral presentation is often the genuine basis for decision making*. Will Marti and her colleagues use the next five days wisely? They will, if they heed advice included in this chapter.

First, we discuss one overall problem that is often termed stage fright, nervousness, shyness, and, more recently, communication apprehension. In whatever guise, fear keeps many talented people from achieving recognition and success that is rightfully theirs. If Marti and her co-workers are immobilized by nervousness at a face-to-face presentation and fail to use the next five days efficiently, their earlier good work might be wasted. The impact of an organizational report resides both in its content and in its presentation. Second, we look at oral communication competence in two very common organizational contexts: (1) interviews with knowledgeable people from whom you elicit the information necessary for an effective written or oral presentation; and (2) structured individual and group presentations in which you explain information and justify interpretations for decision makers.

Nervousness Is Normal

Imagine that you are in the conference room to hear Francine Marti's team present their recommendations. Although facts and figures are very important, you will also find yourself evaluating how strongly the team feels about its recommendations and how confident they are about their projections. If they say they are "certain" about their conclusions, but behave tentatively and unassertively, most audiences would not take their verbal claim at face value. For example, presenters *acting* tentatively usually encourage an audience to assume that messages should be *taken* tentatively. The most reliable clues for evaluating feelings and moods are contained in the *nonverbal* communication realm.

One expert in nonverbal interaction estimates that most impressions of personality and effectiveness are gained by listeners *within the first four minutes* of meeting someone.[2] This situation shocks some people when they hear it. It isn't fair, they think; surely we give each other more slack than that. The answer is that at the conscious level, we often give plenty of opportunities, but at the unconscious level, we reach conclusions rapidly on the basis of nonverbal cues, conclusions that are difficult to revise. Therefore, speakers like Marti must be aware that at the time when she and her colleagues are likely to be most nervous and exhibit the most obvious symptoms of apprehension, the audience is most primed to notice these symptoms. In addition, symptoms of nervousness are often interpreted to mean that speakers are not confident about their content, do not feel strongly about their ideas, or (perhaps worse) are not prepared.

However, effective speakers can anticipate these potential hot spots of presentations and sidestep many of them. As with most basic competencies, two elementary considerations are essential to improving presentational confidence: what to know, and what to do. Because entire books[3] are available to instruct you on how to counteract stage fright, we can only skim the surface here. However, consider the following checklist, containing suggestions supported by years of practical experience and research into public speaking.

WHAT TO KNOW ABOUT NERVOUSNESS

Nervousness is normal and healthy. When you are nervous, it often seems as if anyone else would be perfectly comfortable if they were standing in your shoes. However, this is usually untrue. Nervousness is widespread, natural, and healthy; it is your signal that you are about to do something important. Think about the tasks of your life—such as ordering lunch at a café, choosing dishwashing soap at the grocery store, or driving your normal route to work—and you will realize that you are not nervous doing them. When you feel nervous, it's probably because how people speak and listen face to face can have an impact on their decisions and maybe on their lives. If managed properly, nervousness energizes speakers and raises their alertness and performance to the demands of each new challenge. The key phrase is "if managed properly."

Nervousness can be managed. You want the optimum amount of nervous energy for your presentation. Too much, and listeners might believe you are incompetent. Too little, and listeners might consider you unenthusiastic or uncommitted. Research shows that speakers can control many of the ways nervousness is signaled to others. We will consider several simple reminders and techniques later in this chapter.

Negative effects of nervousness diminish with practice and confidence. Once you feel in command of your material and have rehearsed a presentation thoroughly, and once you have anticipated the range of possible reactions from your audience, you may experience less apprehension. At the very least, your apprehension will be of the energizing variety, easily channeled into the enthusiasm any presentation needs.

Nervousness varies with self-consciousness. Many speakers suffer from the "How Am I Doing?" syndrome. Instead of concentrating on their enthusiasm for their subject matter and how well the audience is understanding their presentation, they focus their fears almost entirely on their own performance—*as* a performance. Effective speakers realize that their subjects and audiences are much more important in the grand scheme of things than their own personal images. Ironically, these persons usually emerge from their presentations with solid reputations, precisely because "building an image" wasn't their all-consuming worry.

Nervousness is rarely as obvious to an audience as it is to an apprehensive speaker. Unfortunately, many presenters erroneously use their internal jitters as a signal of their imagined external failures. In other words, your knees may shake a bit, and you feel a quaver in your voice that isn't normally there. Your hands feel clammy, and there's a bead of sweat on your forehead. You fantasize that the audience *must* be focusing on these things and are ready to dismiss you as a fool or a beginner. Remind yourself that in almost all speaking situations, such signs will be either unnoticeable or inconsequential to audiences. If *you* focus on them, you only distract yourself from your genuine goals— and increase the chances of making mistakes that will create discomfort for everyone.

Competence involves what to know and what to do. If you understand the basic nature of nervousness, you are ready to practice some behavioral changes to manage and channel what nervousness you do experience. One caveat: It's always wise to be a little skeptical when someone else offers you techniques for doing something. Not all techniques will work for everyone, of course. Although these have proved their worth with a variety of professional presenters, you'll have to experiment with them to personalize what works for you.

WHAT TO DO ABOUT NERVOUSNESS

Prepare thoroughly. Many speakers mistakenly believe they're hampered by nervousness when in fact they're hampered by their own lack of preparation. The nervousness they feel is justified; they should be scared. Without solid preparation, they've set up the conditions for their own failure. Adequate topic research, audience analysis, and goal setting can remove or diminish this type of unnecessary problem. Focus on your content and audience. Although a full description of prepresentation strategies is beyond our scope in this chapter, you might begin with the following preparation checklist:

- What is my core message? Can I form the key message into a brief and direct sentence I want an audience to retain after I speak? (If not, you're not prepared yet.)
- Have I done sufficient backup research to be able to explain and, if necessary, defend my core message? (If not, you're not prepared yet.)
- Am I aware of the opinions, attitudes, ideas, and values of my audience, and the specialized language with which they express them? (If not, you're not prepared yet.)
- Do I have specific, relevant, and interesting examples that will help establish the context for the presentation and link my core message to this audience? (If not, you're not prepared yet.)
- Have I clarified for myself a purpose statement describing what I would like the audience to do differently after listening to my core message? Have I planned a behavioral goal? (If not, you're not prepared yet.)

STRATEGIES TO MANAGE NERVOUSNESS IN ORAL PRESENTATIONS

WHAT TO KNOW

1. Nervousness is normal and healthy.
2. Nervousness can be managed by cognitive reminders and behavioral techniques.
3. Nervousness is reduced over time with practice and confidence.
4. Nervousness varies with self-consciousness.
5. Nervousness is rarely as obvious to an audience as it is to an apprehensive speaker.

WHAT TO DO

1. Prepare thoroughly.
   —Know your core message.
   —Research thoroughly enough to be able to explain and defend the core message.
   —Be aware of opinions, attitudes, ideas, and values of your audience.
   —Develop examples to link the core message to your audience.
   —Know what you want the audience to do or think after they listen to you.
   —Practice an opening that presents purpose and context in an involving way.
   —Practice a closing that ties things together and motivates the audience to act.
2. Rehearse your content and style.
3. Dissipate unnecessary energy before the presentation.
4. Converse with the audience, rather than trying to become an orator.
5. Avoid advertising what nervousness you do feel.
6. Involve your audience directly with your presentation.
   —Maintain eye contact with listeners.
   —Use involving language to help listeners visualize contexts and situations.
   —Diversify your presentation with different presentational styles.
7. Just before rising to speak, focus on your breathing, not your text.

- Have I planned and practiced a brief but involving opening that states my purpose, previews the presentation, and clarifies the context for my audience? (If not, you're not prepared yet.)
- Have I planned and practiced a brief but motivating closing that summarizes my overall presentation and clarifies possible action for my audience? (If not, you're not prepared yet.)

Rehearse your content and style. Ask a friend to listen to, and critique, portions of the presentation. However, the simple advice to rehearse applies to all presentations and applies even to presenters who have no one around to listen and provide feedback. If necessary, lock yourself in a room and practice by talking to yourself—even if this feels weird; a lifeless audience response during the real thing feels even more weird. If possible, use a video camera to record your practice session. The more you practice the presentation, the more you are teaching yourself about alternative phrasing, what sounds especially vivid or motivating, and what could make more sense to your listeners. Something that looked good on paper might not sound effective when spoken; as we've pointed out in previous chapters, there are subtle differences between written and spoken language.

It is usually less helpful to rehearse the whole presentation all the way through than to divide it into manageable bits and practice those. As you learn to control the presentation's various sections individually, they will seem more like pieces that fit together to form a coherent picture, rather than an unassembled jigsaw puzzle where all pieces are there but just do not seem to fit. To follow up your prepresentation preparation described above, rehearse in particular detail your opening and closing, along with transitions or bridges between your main points. Despite the importance of facts and figures to accountants, most presentations lose and confuse audiences because of speaker difficulties in beginning, concluding, and linking. In all communication contexts, transitions are important for maintaining an audience's interest, but they are also important for increasing speakers' confidence that they are in charge of their own presentation.

Dissipate unnecessary energy before the presentation. Many speakers are both excited and nervous about an upcoming presentation, and the adrenaline starts to flow. High-energy nervousness, however, can cause fidgeting, pacing, inappropriate gestures, and rapid talk that will not resemble your usual speech patterns. Many speakers find that if they build into their schedules some appropriate exercise before the presentation, their energies are freed toward appropriate enthusiasm, not unnatural and awkward-appearing nervous tension. For example, if Francine Marti is a distance runner, she may be tempted to forgo her normal early morning run and use that time to cram a few extra facts into her presentation. However, her customary five miles may be more helpful in the long run, if the exercise helps her feel more relaxed.

Converse with the audience, rather than trying to become an orator. Try to speak with—rather than lecture at, or read to—listeners. A conversational style will not only be more palatable to business audiences but also help you reframe the presentation mentally from a *trial* or *test* (as many nervous speakers are likely to see it) to a *talk*. This informalizes your task somewhat and encourages you to be extemporaneous. *Extemporaneous* speech is both carefully prepared and informally flexible. Knowing you do not have to have every word planned exactly should remove some emotional pressure you may feel. Audiences listening to extemporaneous speakers rarely judge them harshly when they make verbal slips, and recognizing this also removes some pressure. While there are times when reading from a manuscript is expected (for example, legal language or formal reports to stockholders), in most cases you should be working only from a set of notes. If you have the same presentation to give several times, maintaining a conversational tone means that you might change phrases or add new illustrations as they occur to you. In developing your own conversational style, make sure that you remain suitably professional for the circumstances in which you find yourself.

Conversational presenters sometimes are tempted to substitute silliness or giddiness for substance, an approach that only trivializes the presentation. Adverse audience reaction to overly informal speakers can actually increase everyone's nervousness level during the presentation, creating the exact problem you are trying to avoid. Effective speakers still are organized, even while conversing with an audience. Take a detailed outline to the front of the room, even if you think you will not need it. It can be your parachute back to solid ground if you find yourself losing your organization or focus.

Avoid advertising what nervousness you do feel. Francine Marti might be tempted to say something like, " Wow, I have hardly ever done this, so bear with me. My knees are really shaking!" Although many listeners will certainly sympathize, such a comment does nothing to improve her credibility. Audiences may like some presenters enormously, but they usually make bottom-line business decisions on other grounds. They want to be confident about your recommendations, and your disclaimers undercut their confidence in you. Disclaimers and apologies inhibit listeners from reaching the same solid conclusions you are advocating. A second problem with advertising your nervousness, of course, is that you risk focusing your own attention on it more than necessary.

Involve your audience actively in the presentation. When people vegetate as uninvolved listeners, their attention naturally wanders from your content to doodling, grocery lists, judging your apparel, nitpicking small points, and other peripheral matters. As you observe these distracted behaviors, you will start to doubt yourself, thus becoming more nervous. Keeping your audience at least minimally involved is not a guaranteed antidote but will usually help control your jitters. Consider these suggestions for low-level audience involvement:

- Maintain eye contact with as many listeners as possible. They will appreciate this demonstration of your connection with them, and people who have your attention are more likely to give you theirs. This means engaging people individually rather than sweeping your eyes over the room. Further, forget the advice you may have heard about looking at a spot on the back wall over the eye level of your audience. This tactic, which has doubtful merit even in large auditoriums, is downright silly in the vast majority of business presentations, which are conducted in a conference room or medium-sized office. People readily can see where you're looking.

- Use psychologically involving *action language,* which requires listeners to dramatize choices, compare options, visualize situations, and place themselves in hypothetical or real contexts. For example, you might say, "Imagine yourselves in this same room in nine months, debating possible expansion. Will closing the Dayton plant now be a boon or a hindrance to those discussions? That's the kind of question we're facing here." Debating. Closing. Facing. This kind of action language makes audiences consider options, and frames them as participants within *your message* rather than evaluators of *you.*

- Diversify your presentation with different styles and modes. For example, in a team presentation, one person could begin with a low-key PowerPoint demonstration of recent investment trends. Then a speaker with a more enthusiastic verbal style might summarize findings and recommendations, with the first speaker finishing up, distributing a handout on the presentation, and inviting audience questions addressed to either presenter. An individual presenter could implement the same strategy by using multimedia, vocal variety, or other audience involvement techniques like hypothetical questions.

Before rising to speak, focus on your breathing, not your text.
You do not need to fret about your message or the audience reaction right up to the moment you start to talk. Most professional speakers control excessive nervousness by small behavioral tricks or rituals they do just before rising to address an audience. For example, as you silently wait for your time to speak, ensure you are seated in a comfortable and stable posture. Silently and unobtrusively, draw in 5 or 10 breaths, imagining with each one the route of the air as it descends to the center of your body at the diaphragm. Hold the breath for a second after each intake and then exhale gently. Draw in and release the breaths in a calm, measured manner, thinking only about counting them. The counting takes your mind off the nervous tension you might be feeling, while the centering breaths counteract the tendency toward shallow breathing brought on by anxiety. Most speakers find that their muscles relax, their minds become more flexible and creative, and perhaps, most important, fears associated with communicating with an audience begin to ebb.

Interviews to Gather Information

Your ability to make a persuasive and clear presentation to a group will depend upon a variety of factors. One obvious facet of success in an organization is the extent to which you are informed and prepared. Some information-gathering strategies depend upon your ability to do library research or extended personal research in on-line databases or organizational files, such as we discussed in Chapter 4. Still other research strategies depend upon oral communication skills in face-to-face communication encounters. This is another crucial facet of the connection between accounting and communication.

Full preparation for presentations will often involve interviews with key people both inside and outside your organization, eliciting the information that could

Analyzing Normal and Abnormal Nervousness

Recall the past three times you have spoken before an audience of four or more people. For each event, do the following tasks.

Tasks:

1. Describe as many details about your own behaviors and others' reactions as you remember.
2. Describe all the symptoms of nervousness you can remember (rapid speech? dry mouth? quivering hands? running out of breath? forgetting your organization?).
3. Analyze possible links between your nervousness levels and the unique circumstances in each situation. For example, were you more or less nervous when using visual aids? when speaking to a group of 20 rather than a group of 4 seated around a small conference table? when you had a large amount of information to summarize, or when you wanted to persuade the audience to take a certain specific action?
4. How might you have revised the presentations (or your preparation for them) using the suggestions of this section? Which approaches seem most applicable and valuable to you? Which seem less applicable and valuable?

become the very foundation for your best idea. Effective interviewers understand that information can take many forms. Some interviews will be attempts to determine the facts of a situation ("Who signs checks greater than $1,000?" or "In what form do you keep records from prior years?"); other interviews will ask for critical judgments from your respondents ("What did you think about Chrysler's decision to close the Austin plant rather than retool for the Caravan?"); while still others will probe the interviewee's feelings and personal preferences ("Would you be more comfortable with calling in outside consultants, or would you rather decide on the new software in-house?"). Interviews can also have political purposes and implications. If you are going to make suggestions about someone else's turf, interviewing that person can serve several functions beyond simply providing information: It can generate critical support, identify problems, demonstrate your own expertise, and be the professionally courteous thing to do.

Try to think of the term *interview* without the connotations of nervousness we attach to job interviews, and without the connotations of interrogation we associate with an IRS agent interviewing you about last year's income tax filing. Instead, consider an organizational interview as literally an inter-view, a focused interpersonal conversation in which one person elicits information from another or others, yet in which each party gains some knowledge ("view") of the other through the relationship ("inter"). Although a comprehensive treatment of interviewing approaches is impossible here,[4] we will stress four major issues: preparing, questioning, listening, and recording.

PREPARING FOR THE INTERVIEW

Unfortunately, an interview occasionally is conducted for purely selfish reasons. The interviewer could find out information on his or her own but would rather shortcut the process a bit by calling Carl or Carla and asking for a date to have lunch. During the hour of lunch, sometime near the second cup of coffee, comes the hidden agenda: "Oh, by the way, Carla, could you fill me in on why the Mimico acquisition fell through? I've got a report to do for the home office. What was their financial baggage, anyway?" It is probably obvious the questioner does not have enough background in the situation and is substituting this meeting for more systematic information gathering that could be done alone. The respondent, in effect, is asked to do some of the legwork for the interviewer. If you value your organizational reputation, avoid this kind of manipulation that masquerades as sincere interviewing.

Many interviews, on the other hand, are absolutely necessary. Information you might discover in computer files, in company correspondence, and in the library will be incomplete without the context that key organizational sources can provide. *Context,* as we have stressed in Chapter 4, is what surrounds an event, often determining how it should be interpreted. Context can include such factors as the history of a gradually emerging decision, the kind of time pressures in which a decision was made, or the interpersonal atmosphere of cooperation or disagreement from which an edict issued. In addition to the obvious context factors, many organizational facts are not easily researched in the files simply because they are not quantitative or even objective. They may be personal, value- or emotion-based, culture-specific within a given department or company, or oriented to future goals that may not yet be articulated precisely. All these factors define the human side of organizational life; they are the territory of the interview.

Be sure to do your homework before any interview. Not only should you familiarize yourself with relevant facts, but you should also make sure you have read the key documents your respondent is likely to discuss, and that you are able to speak his or her language as much as possible. Some subgroups or jobs develop their own distinct *jargon,* their own specialized terminology for their distinct tasks and challenges. Some jargon terms may be linguistic shortcuts, in which a common or seemingly familiar word is used in a particularized meaning. Your company's legal department, for example, will prepare "briefs" that are anything but brief. Other jargon words may sound esoteric and unfamiliar if you do not work within a given group, but are used so frequently within that office that an interviewee usually will not think twice about using them in a conversation with you. In such circumstances, it is your responsibility to prepare yourself beforehand to converse as much as possible within the other person's field of information and language style. However, never be afraid to ask for clarification within interviews if you have the slightest doubt about terminology or context. Remember that precision, clarity, and specificity are the hallmarks of your profession; employ them in your interviewing practices, too.

QUESTIONING AND THE INTERVIEW SCHEDULE

"What do I want to know?" is the question interviewers have to answer before deciding what to ask. Adequate preparation gives you the answer to this basic orienting question of an interview and can lead to the specific questions you will want to ask your respondent later. In essence, the answer to this question signals the theme of your interview, just as each report you write should have a major purpose or theme that it develops. Although there will obviously be times when what you need to know is fuzzy and the interview will have to be somewhat exploratory, in most business settings interviews are expected to have a clear focus. The interviewer is expected to propel the interview forward with appropriate and well-chosen questions that are built upon what the respondent actually says.

This advice, to let the interview be structured largely by the interviewee's talk, seems to contradict what novice interviewers believe an interview should be. They naively think that an interview is a tidy sequence of preset questions that primarily build on each other, that lead to a rational point or the possibility of a logical conclusion. This is a misconception that derails more interviews than it assists. In contrast, the interviewer's most essential task, in effect, is assisted talk. He or she wants to help the other person elaborate on information, explain motivations, and provide context. This rarely happens as a result of predecided questions, but it often develops within an interchange that has a conversational quality. The questions do not necessarily build on each other all the time, but they should build on the responses. What does this mean for the interviewer preparing to meet an interviewee?

The interview schedule. Our advice is to prepare a tentative *interview schedule* based on what you need to learn from the meeting. The schedule is a short and coherent list of questions (often five to eight in brief interviews) on issues you assume are important. The order of questioning may be chronological (earliest to latest), cause-effect (what events led to what outcomes), problem-solution (inquiries about difficulties setting the stage for inquiries about ways to resolve them), or some other pattern of your choice that would be meaningful to the interviewee and appropriate to the content. The crucial thing is that you should have a pattern; often the interviewees in business contexts are equal to or greater in status than the interviewer, and you do not want peers and superiors to believe you're disorganized.

The main value of a schedule is to serve as the spine of the interview, its central support if things seem to be tottering. However, it is dangerous to assume that the asking and answering of such questions constitutes the whole interview. Be flexible and prepared to redirect the interview toward side trips of meaning that may be suggested within the conversation. If you are overly committed to simply going down a list of questions, you will probably miss important information. Be careful to listen and reflect before going on. In addition, avoid planning your next question while the person you're interviewing is still responding to your current question.

Probe questions. Based on how well you listen to your respondent, you should be reminded of secondary questions designed to follow up or clarify your interviewee's statements. These follow-ups, usually called *probe questions* by experienced interviewers, may not even be questions in the usual sense. For example, you might probe with such comments as: "You're concerned about receivables" or "I'm noticing that when you talked about the Pepsi contract, you mentioned several projections, but just now, on the Dr. Pepper deal, the matter seems less defined." A question may be implied by the probe, but it's one that the respondent may choose to answer in a variety of ways. Other examples of probing styles:

> "Seems to me that you think the company's 1999 responses were stronger than those of earlier years."
> "That's an interesting point. Could you say more about its tax implications?"
> "I'm not sure I understand why the client's account was closed. Did the controller have any information about that company?"

The basic goals of probing are (1) to satisfy your spontaneous curiosity while interpreting a respondent's answer; (2) to ensure you understand a response—including the assumptions behind it—before moving on to the next question; and (3) to invite an interviewee to provide more details, examples, or illustrations. Many interviewers train themselves to conclude each interview with an "Anything else?" question that functions as a wrap-up probe. Interviewing texts typically call this a *clearinghouse probe*. For example, finish with something like:

> "Thank you so much. I think this has been very helpful, and I learned what I needed to know. Before I leave, though, I'm wondering if there's anything else you want to tell me about [the topic]. Anything you think I might have missed or forgotten to ask?"

You'd be surprised how much relevant information comes to mind when the pressure of being questioned is released a bit.

Implications for relationships. Questions are indispensable in interviewing, but they have implications for relationships. Two issues are especially important to consider in the organizational context. First, asking a question might suggest in some circumstances that the interviewer is deficient, ignorant in a given area, and in need of the help of a more experienced, higher status person. This is especially true if a question is asked in a subservient tone, or asked in a reticent, retiring, or tentative way. Most psychological states (e.g., fear or insecurity) are signaled by messages of voice (e.g., pitch or volume), posture, gesture, and eye contact. A too-soft, quavering voice coming from an interviewer with averted eyes is unlikely to impress corporate interviewees. Therefore, be aware that how you ask certain questions, and even the act of questioning itself, can undercut your status and position within an organization. It is one thing to be appropriately

---

QUESTIONING IN AN INTERVIEW

1. Ask about what you want to know. This sets the theme of an interview.
2. Recognize that interviewers are expected to propel the interview.
3. Remember that an interviewer's basic task is to help others talk.
4. Prepare a tentative *interview schedule* based on what you need to learn.
  —Provides spine of interview.
  —Contains a short and coherent list of questions or topics.
  —Allows for recommended sequences: chronological, cause-effect, problem-solution.
5. Be flexible and listen.
  —Reflect before going on.
  —Develop probe questions designed to follow up or clarify the interviewee's points.
  —Ensure you understand a response.
  —Invite additional details, illustrations, examples.
6. Be assertive—establish a conversational relationship with the interviewee.

---

respectful of a superior, for instance, but quite another thing to advertise an inappropriately deferential self-concept. Consider your questioning style carefully, and ask questions assertively.

Second, questioning someone could at times suggest the opposite condition, an overly aggressive power message. Questioning can take on the tone of interrogation, implying that the person questioned is of lower status and on the spot. Often, as in a courtroom or classroom, people will not ask questions because they need or want answers but because they are trying to test someone else's ability to answer. In daily life, too, people often use questions sarcastically to substitute for negative statements that they are reluctant to make openly. This is why being questioned often stimulates defensiveness. Even such seemingly innocent questions as "Why did you do that?" in some contexts might be interpreted by another person as "really" meaning "You should have done something else," "That was stupid," or "You should have asked someone else about it before you decided things on your own." Again, most of this kind of meaning is implied by your nonverbal demeanor and vocal tone. In the successful interview, you neither undercut your own status (whatever it may be) nor undercut the position of the other. You talk with the other person about issues of mutual concern.

LISTENING

Your interviewing skill will depend largely on your listening skill. Unfortunately, most executives assume that while they may need training in public speaking, listening is somehow an automatic human activity, much like breathing. Yet it too should receive increased attention and skill practice, because it is the other half of improved oral communication. It is a crucial skill in all interpersonal situations, but when you are trying to learn by interviewing others, listening is especially key. Our basic suggestions for improving listening should help your interviews progress more smoothly. You may want to consult other specialized sources for more information on this vital communication skill.[5]

By listening carefully, you will learn more than by simply recording replies robotically to canned questions you'd planned before ever walking into the room or placing the telephone call. Although most people assume listening is a simple process of taking in or receiving oral messages, it is considerably more complicated than that. Listeners not only *hear* messages (respond at the physiological level to sound waves) but inevitably *interpret* them as well. The process of *listening* involves integrating new oral messages into the patterns of meaning the listener has already developed. Whereas hearing is a biological process of registering sounds, listening can be thought of as a psychological and social process of meaning-making. This definition suggests a series of practical reminders:

When listening, you will be prejudiced. You must learn to recognize and take into account your own biases, since psychologists tell us that it's impossible for you to check them at the door. The decision to listen to someone means that your biases are already at work, perhaps distorting what the interviewee says to fit better into what you already believe or expected to hear in the first place. Alternatively, your bias as a listener may at times exaggerate the contrast between your perception of the speaker's meaning and what you believe about a topic. In any case, no listener can receive messages in a totally objective, unprejudiced way. New messages are inevitably related to old patterns and prejudgments. You cannot assume that your interpretation is what the other person means. It would be safer to assume that what he or she means is *never* precisely what you've interpreted.

Your listening style will be observed and evaluated by speakers. It's equally dangerous to assume that just because listening is silent thought, your listening won't be obvious to your interviewee. People tend to signal their listening to companions by displaying certain behaviors. In Western culture, for example, listeners tend, with certain exceptions, to maintain consistent but not constant eye contact with speakers, orient their posture generally toward speakers, confirm that speakers have been heard by providing such conversational encouragers as nodding appropriately or by saying "uh-huh" occasionally, avoid subsidiary tasks (such as reading a newspaper while listening) and imposing nonverbal barriers (such as wearing dark glasses), and participate periodically by contributing to the flow of the conversation. In the absence of such behavioral evidence, speakers tend to assume that listening is not occurring effectively. Thus, your listening style, especially aspects of it of which you are not fully aware, will often directly determine the amount and kind of information others are willing to share. If they believe you to be a poor listener, someone who doesn't care about them or their ideas, or someone who wouldn't understand a nuanced idea, they'll curtail their talk with you.

When listening effectively, you identify with and empathize with speakers. If listening naturally involves your own biases, then effective listening must somehow invite meanings that go beyond your prior experience. Listening involves learning other perspectives. This happens through *empathy,* the willingness to imagine another person's experience from his or her perspective while remembering that you cannot give up your own perspective. Although some training programs teach empathy as a behavioral technique, we prefer to think of it as an *attitude of communication* by which you remind yourself that your interviewee's perspective is both important and separate from your own. Empathizing is different from sympathizing; sympathy is associated with feelings for the other person (often in a sad or negative situation), whereas empathy is a more generalized willingness to try to experience the world as other people do from the standpoint of their experience. Remember that you don't have to agree with, or even like, a person in order to empathize.

When listening effectively, you check often on the other person's meanings. This approach, often called *active listening* by communication trainers, is essential if you want to convince others that you care about what they say. Listening actively should reflect your genuine interest in people and what they think. If you are in fact understanding someone's message well, demonstrate this occasionally by paraphrasing their ideas. You are not merely trying to repeat their words back to them as a parrot or a tape recorder would but instead are demonstrating how you are interpreting their ideas. At times, you might even want to go beyond what they've said overtly to check out a new impression you have started to develop without much evidence. As examples, active listeners often use such comments and internal summaries as:

> "Your main point seems to be . . . "
> "I think what you're getting at is . . . "
> "I think you're saying that you've got two responses to the merger—that you're ambivalent, and that . . . "
> "Somehow, today you seemed more pleased by this tax interpretation than when we talked last month. Am I wrong?"

LISTENING IN AN INTERVIEW

- Understand your own prejudices.
- Recognize how your listening style will be observed and evaluated by the speaker.
- Identify and empathize with the speaker.
- Check often on the speaker's meaning, using *active listening.*

REMEMBERING AND RECORDING

No matter how well you listen in an interview, you will not remember everything. The decisions you make about how to record and recall information will directly affect the value of the interviewing. Actually, these decisions also significantly affect how the interviewee will interact with you during the conversation. We will survey three major options for retaining interview information: note taking, delayed note taking, and tape recording.

First, the most traditional and time-tested method for recording information is to take extensive notes. The advantage of *note taking* is that it is relatively unobtrusive and combines the ability to be thorough with the possibility of devising your own system of abbreviation or shorthand. Unfortunately, the more extensive your note taking becomes, the less you can maintain visual contact with, and nonverbal sensitivity to, your conversation partner. While taking notes, we suggest that you:

- Be as unobtrusive as possible in your writing. For example, shuffling stacks of papers will usually distract both parties.
- Take notes only (but always) when the respondent discusses an important and germane point.
- Key your notes to the interview schedule you have prepared beforehand.

The planned interview topics or questions might be jotted down on the left-hand side of a page (or each at the top of a series of pages in a small notebook) with plenty of space to the right of each topic for noting the other's responses.

A second approach preferred by many experienced interviewers could be called *delayed note taking*. In this method, you postpone written reactions until immediately after the interview. The key to delayed note taking is to conduct the interview with maximum alertness and then set aside about 30 minutes after the interview itself to reconstruct on paper the key information from the interview. At this time, review the schedule of questions and topics, jotting down what significant replies the respondent made. Also, note the tone of the interviewee's talk, if relevant, and any specific documents or sources that were suggested to you during the interview. Sometimes you find that documents have been misplaced; mentioning them in your notes can help solve this potential mystery.

Even if you plan to use a delayed approach to note taking, you should carry a notepad into an interview, if only to be prepared for a situation where the respondent asks or expects you to remember something specifically. Your notepad, even if you use it sparingly, can be reassuring to interviewees in a professional context. Of course, there is no reason why you couldn't combine judicious note taking within the interview with a private review session afterward; this combined strategy can only increase your thoroughness.

A third option should be reserved for the times when exactness of detail is your foremost criterion. Consider the possibility of tape recording an interview, preferably with a microcassette recorder or other small and unobtrusive cassette recorder placed slightly to the side of the conversation. Ask first if your respon-

> ## REMEMBERING AND RECORDING INFORMATION IN AN INTERVIEW
>
> - **Note taking:** Key on responses to questions and topics from your interview schedule.
> - **Delayed note taking:** Set aside 30 minutes after the interview to write intensively about impressions and record details.
> - **Tape recording:** If the interviewee is not bothered by taping, use a small cassette recorder off to the side of the conversation. Be sure to familiarize yourself with the functions of the recorder, and ensure you have fresh batteries.

dent is comfortable with the recorder, and then, even with agreement, be sensitive to signals that might indicate discomfort (fidgeting, constant glances toward the machine, comments about the taping process). If you have doubts about the appropriateness of taping, do not use it. The extra information that some public sphere interviewers such as reporters get from taping an interview is usually not worth the trade-off you will make in an informal organizational context. For example, a reporter may need to enliven a story with quotations, but for most organizational research purposes, context and basic factual information will be much more crucial than a respondent's exact wording.

In sum, the process of interviewing can be seen as the skill of focused conversation. It is a conversation in which one person approaches the interaction with a need for information or insight that another presumably possesses, and a conversation in which the interviewer facilitates the interviewee's sharing of that information. It is not a mechanical process but rather a potentially enjoyable way for people to learn about the organization. Still, however enjoyable the process is for participants, practical interviewing must be efficient, effective, and informative. Along with other forms of research we have described in previous chapters, it sets the stage for the kinds of persuasive presentations accountants are called upon to make.

Presentations Are Not Speeches

Many executives mistakenly, and often selfishly, believe that because they are speaking, the audience is primarily interested in them. Perhaps in the best of all possible worlds where friendly people have unlimited time, no deadlines, and maximum interest in each other, they could share many of their personal feelings with their co-workers and expect listeners to be fascinated. Unfortunately, the real world in which most of us operate falls somewhat short of ideal. Listeners at work, though they may like you very much, still must be interested primarily in the instrumental outcomes of your presentation.

In most professional speaking situations, you can probably assume these things about your audience:

Practicing an Interview

Videotape or audiotape an interview of an economist or other financial expert from a program such as PBS's "Wall Street Week."

Tasks:

1. Identify in the tape any potential misunderstandings by the interviewer, or points where viewers could easily misinterpret or be confused about what the expert is saying.

2. For each potential incident of "missed meaning," write out a probe response that would paraphrase the expert in order to check on, clarify, or specify your perception of his or her intended meaning. Remember in doing this that you are not mind-reading the intentions of the expert; an *active listening* response, sometimes called a "perception check," can be just as productive if you turn out to be wrong in your paraphrase. You are giving the other person a chance to clarify things, and the result is better understanding.

---

1. They are busy.

2. On some level, they would rather be somewhere else.

3. They want to spend their time efficiently.

4. You matter to them as a person, but they especially value your capacity to help them understand something they don't currently know.

To help your audience, and therefore to help yourself, plan on giving business presentations rather than speeches.[6] According to one specialist in business communication, *presentations* are speech occasions involving commitments by individuals or groups to focus on a topic in order to help an audience accomplish something important. Presentations are specific and necessarily involve a direct evaluation by the audience of the worth of the speaker's commitment.[7] They are pointed and lean. Speeches, on the other hand, are folksy, friendly, and flabbier. In a speech, a person may talk about himself or herself, about interesting topics in general, about wide-ranging implications of things. If presentations get slimmer as they go along, speeches get fatter.

This is not a simple distinction of good and bad, because communication depends upon context. In most formal communication settings in which accountants find themselves, however, the *efficiency criterion* listeners use to evaluate presentations is more appropriate than the *involvement/interestingness criterion* with which we tend to evaluate speeches. Speeches can explore winding trails of stories and anecdotes in order to involve or even entertain an audience, and this is appropriate for many situations. But most of the business world, fortunately or unfortunately, must rely less upon scenic trails of talk and must travel on more direct highways.

Francine Marti's team cannot afford to violate their audience's expectation of efficiency. The group will probably be allotted only a brief time period to present

THE ROLE OF PRESENTATIONS

MANAGEMENT CONSULTANT THOMAS LEECH:

. . . Top management is using presentations as never before to review studies, requests, recommendations, and programs being undertaken by their employees. Requests for new equipment, go-ahead to pursue a new business line, review of planning for facility modernization, findings of a task report on employee morale—all will probably involve one or more presentations with increasingly higher management. The oral presentation may be the main avenue of communication, and key decisions may be based on the presentation.

Unfortunately, many people . . . treat the presentation lightly. They spend months working on a task, and spend two days on the presentation of it . . . . Over and over again I have seen presenters come to a sudden realization in the middle of an important presentation that they should have spent a bit more time preparing. That "winging it" doesn't work. That it is terribly embarrassing not to be able to answer obvious questions with their bosses present. That it is a bad blunder not to have rehearsed.[8]

and justify their conclusions. They must plan that time carefully to ensure accurate understanding.

The oral presentation may be the main avenue of communication, and key decisions may be based on the presentation. Such opportunities may involve either individual or team effort. Let's consider some of the basics of each type of approach.

INDIVIDUAL PRESENTATIONS

Earlier chapters in this book discuss the importance of developing a communication plan when making an oral presentation. You must analyze the context compared with your own purpose, analyze your audience, decide on the most crucial content for your presentation, create a tentative text, and then polish what you want to say relative to this audience's knowledge and need to hear it.

Beyond these factors, however, you can improve presentations by following some more specific advice:

Plan for spontaneity. This suggestion may sound like a paradox, but it is one born out of the particular challenges of speaking with a live audience. The value of spontaneity is based on two interrelated observations of effective speakers. First, audiences appreciate extemporaneous speaking, speaking that is both planned efficiently and spontaneous enough to take into account the immediate circumstances of the presentation. Second, with experience comes the realization that not everything will run according to plan. Oral presentations, like live television, are particularly susceptible to Murphy's Law—things that *can* go wrong . . . *do* go wrong. Effective speakers, therefore, create contingency plans. For example, when your laptop computer crashes in the middle of your presentation, eliminating the multicolored charts you had prepared so carefully, know the data well enough to describe interrelationships vividly with a verbal explanation rather than relying on visual cues. Part of planning for spontaneity, of course, involves setting up your room and equipment prior to the start of the presentation and knowing where everything is located. Contingency planning also involves preparing backup hard copy handouts or transparencies even if you may be preparing a PowerPoint presentation.

Sensitize yourself to nonverbal cues. The primary distinction between written and oral communication contexts is that talk allows a fuller range of human expression. Even though oral communication content is in many ways easier to plan for, the nonverbal aspects of your presentation will often make or break the occasion. Some communication researchers, in fact, have estimated that 67 to over 90 percent of the meaning humans generate while communicating can come from the interpretation of nonverbal communication messages. A sensitivity to the nonverbal realm is even more critical because, according to psychologists, most people base their judgments about relationships primarily on nonverbal cues, and when nonverbal and verbal behaviors of a speaker contradict each other (for example, a sweating and shaking speaker asserting "I'm confident you'll accept my suggestions"), the listener is likely to put faith in the nonverbal message. Before and during your presentation, therefore, be aware of the following:

- **Dress appropriately for the occasion.** In most situations, you will make your choices based on some knowledge about your audience's likely clothing choices, and you will match those choices. However, at times you may not know this information within your organization, and you might even be asked to speak to members of other groups as well. In such contexts, here's a reasonable guideline: Think about the continuum of normal and acceptable wardrobe choices for employees in your organization and comparable ones; then, dress in the upper range of that continuum for your presentation. Few speakers will ever be criticized for "overdressing" a bit, but many have been embarrassed when they have "underdressed" for a presentation to a room full of people in conservative dark suits.

- **Be sensitive to time constraints.** Arrive on time, and do not keep an audience beyond the allotted or expected time.

- **Speak from a distance appropriate for your audience and room.** Different distances are appropriate for different topics, interaction contexts, and persons. For example, in a large room, audiences will expect presenters to stand at the front at least 10 or 15 feet away, while many conference rooms seem to invite interaction distances of only 4 or 5 feet, and a seated presentation may seem more reasonable.

- **Use your voice to indicate by its tone, rate, volume, and inflection that you are interested and credible in your topic area.**

- **Handle your materials and equipment confidently.** For example, speakers who pass out complicated handouts that introduce ideas not yet covered in the presentation usually succeed only in distracting audiences. If handouts are necessary, match their complexity to the specific point for which they are used, and time their distribution appropriately. If you are using any type of overhead projector or computer program like PowerPoint for illustration, ensure your materials are in order and accessible, and the equipment is in good working order.

- **Look listeners in the eyes.** Few messages indicate power relationships more effectively than eye contact patterns. Without staring, speakers are usually expected to connect visually with all audience members in small group gatherings. Of course, cross-cultural expectations are especially important in this regard, as they are with most nonverbal cues. Make certain you understand the eye contact patterns expected by the groups you are addressing.

- **Develop appropriate animation in your gestures.** Body movement can help your audience relate more directly to you and your commitment to the topic. Appropriate gestures are those matched to your personality, of course, but they also should emphasize important points in your presentation. However, speakers have different styles of animation, and everyone's style will be distinctive. Rather than practice gestures that you think are impressive for others but that may appear too artificial or phony in your talks, our advice is to display *natural* movement while presenting. In reviewing your practice videotapes, make sure your "standup" style of presenting roughly mirrors your conversational style of gesturing. If you are not sure about how your presentational gestures match your informal conversational gestures, ask a friend or two to critique the practice tapes and respond to this point.

Introduce the presentation by clarifying your own credibility. Even if the time allotted to you is brief, do not launch directly into your recommendations and analysis without a short summary of who you are and what kind of work led you to this particular presentation. Without boasting, describe factually (*a*) your qualifications, if not known already by your audience, (*b*) the charge, or what you were asked to do or recommend; and (*c*) the extent of the research and preparation that led to the presentation.

For example, an accountant might say: "As some of you know, six months ago I was asked by our administrative vice president to investigate payroll procedures across all our subsidiaries. As you might also know, I worked on payroll matters when I was in public accounting. I spent about three months gathering data by collecting written reports and conducting face-to-face interviews, and another two months or so analyzing the procedures and investigating what similar firms are doing in streamlining payroll matters. My assistant, Fred Smith, and I wrote the report you received last week—and I'm here today to summarize the recommendations and answer any questions you may have."

Preview the presentation for the audience, including how much time you believe it will take. As we discuss in Chapter 4, audiences want to know your purpose and the context in which they are listening. It's especially important for listeners to be given a mental map of where you are taking them: "After I share some data about wasted resources that may surprise you, I'll describe three basic recommendations for changes in procedures. Finally, I'll take whatever questions you may have." Many presenters will forget the preview step because they are nervous, launching directly into point one (or, worse, point two or three). When this happens, audiences tend to get nervous. Especially in a business environment, people like to be assured that the person nominally in control of the time of a meeting (usually the speaker; in this case, you) understands the time constraints and has a realistic grasp of how information fits into those constraints. At this point, present your listeners with your hook, the basic reason that should motivate them to continue listening: "Adopting the procedural changes I propose could save the firm several thousand dollars and hundreds of work hours per pay period."

Stay with your planned organization unless audience reaction persuades you to deviate from it. Although most content- and organization-related issues of oral presentations have been discussed in previous chapters, this point deserves special emphasis here. A truly extemporaneous style means that your presentation is connected both to the content plan you have already devised and to the audience whose reactions you cannot fully predict. However, what should happen when, during the early stages of the presentation, you sense, perhaps by comments or nonverbal reactions, that the audience already fully understood two of your three recommendations even before you began? Should discussion time *still* be divided evenly among the recommendations, as you'd planned? Although the answer seems obvious and logical, many inexperienced presenters unfortunately stick with their canned outline, no matter what. The audience suffers, as does the cause being advocated.

Review for impact. Most presentations are short, so there is little time for elaborate closings. Still, you should include a brief review that summarizes your main points and gives the listeners a strong reason for remembering what you have advocated. For example: "In the past 15 minutes, you've heard three reasons why we might want to revise our payroll procedures. The first two reasons are direct bottom-line factors: streamlining office record keeping will save money, and combining the responsibilities of several accountants will save both money and time. The final reason is basic fairness: Our employees will see us as more concerned with their welfare."

Stress dialogue. Many speakers mistakenly think presentations are one-way pipelines to an audience, similar to written communication. However, if this is the prime purpose, why arrange for the direct presence of an audience at all? People gather in groups to do more than just take in information; that can be done more efficiently through other communication modes. People come together, rather, to test ideas creatively, to compare their judgments with others, to clarify ambiguities, and, sometimes frankly, to demonstrate directly that they know what they are talking about. Oral presentations, in other words, are social occasions. Don't treat them as mere one-way performances.

The basic reservation to this advice, though, is that dialogue cannot be allowed to take over the presentation at the expense of the core message and the evidence that supports it. At times, you must exert some control to limit dialogue, in order to be able to present the full range of information you had planned. In general, it's best to structure a time for reactions and discussion after the content foundation has been laid. Therefore, when concluding the presentation, be sure to ask for reactions and welcome them when they come. Usually you will be the person in the room who knows the most about your specific topic, even if you won't necessarily be the one with the most power. In Chapter 6 we suggest that you deal with questions and replies through a simple three-stage approach:

- First, *restate or clarify* the question to the questioner's satisfaction, to prove that you understand. This is similar to the active listening process we discussed in the preceding section on interviewing. In larger audiences a restatement also serves to specify the question for audience members who may have missed part of the questioner's intent or who have not heard it all. For example: "You're wondering if our experience last year is typical of the industry."

- Second, *answer* the question, preferably by providing new information and by relating that information to earlier content from the talk. For example: "Well, we're very close to the industry mean on this measure . . . actually, just a little above it. But we now lead the companies from the Midwest. As recently as 1990, the shipping advantages of most Midwest firms made them dominant, but now we seem to be the only one flourishing."

- Third, *check back* to verify that the questioner feels at least understood, if not agreed with. For example: "Did that answer your question?"

Although each question-and-answer period will be different, several additional guidelines will help you manage this crucial and often tense section of your presentation. Many presenters who do a great job planning and explaining basic information to audiences find themselves clutching up when asked relatively simple questions.

- **Leave plenty of time for reactions.** The more complex the information, and the more difficult the subsequent decision making, the longer you'll want to allow for questions. Given highly complex matters, a typical rule of thumb is for one-third of the total presentation time to be set aside for a question-and-answer period. For most topics, plan for somewhat less time.

- **Practice a way to invite questions sincerely.** You do not want it to appear that you are inviting questions just because you have run out of things to say or, on the other hand, that you will regard questions as attacks on your credibility. "Well, that's about it. I don't know if you have questions or not, but I could spend some time clarifying things here," is not a terribly productive comment. Instead, think about something more inviting, like: "In our limited time, I've covered a lot of ground but had to move rapidly. I wouldn't be surprised if there were quite a few questions and reactions. We have five or six minutes left, and I'd enjoy the opportunity to address them."

- **Avoid commenting on the quality of individual questions.** This includes replying with compliments, in-jokes, or (especially) sarcasm. Even if you think you know why someone is asking a question, or even if you think they should already know the answer, respond succinctly and clearly instead of highlighting how unnecessary, wonderful, silly, or trivial the question might be (in your opinion). Let others draw the inferences, positive or negative. (Remember too that when you tell someone how great a question is, which seems like a good thing, subsequent questioners notice when you do not compliment *their* questions; you have set up an implied dimension of judgment without necessarily meaning to do so, and it may be the Big Boss who winds up feeling slighted!)

- **Do not bluff.** If you are sufficiently prepared, as you should be, you will know enough to respond to most questions. However, no one can answer everything. Audiences appreciate the honesty of speakers who say "I don't know" as long as it's not on a topic the speaker clearly *should* know about. However, be sure to tell listeners that you will find out the answer promptly and get back to the questioner (or others) with it.

- **Call on a variety of questioners.** Though some occasions will make it virtually impossible to ignore persistent questioners who hog the time, audiences appreciate your ability to give the floor to as many questioners as possible.

- **Do not get sucked into arguments.** This is a relatively structured interaction, not a free-form conversation. At times, questioners will be asking you implicitly to support their positions vis-à-vis other executives or managers, and you will have no way of knowing the dynamics of those arguments. Other times, you will feel under attack for reasons you cannot fathom, and this will seem unfair. The question-and-answer period in a presentation is a terrible forum to try to make yourself understood in response to such dilemmas. If you feel yourself becoming trapped in these larger issues, insert a diplomatic-but-true suggestion like: "This is obviously a thorny issue. Would it be possible to meet later, when we have more time to air things out? I may not understand yet all the problems you're worried about."

Follow up the presentation with a realistic analysis of feedback.

In some situations, as in presenting workshops or training modules, you will be

Analyzing an Oral Presentation

PRACTICE

Have a friend with a video camera tape your five-minute infor-mal explanation of the major features of your favorite word processing, database, or electronic mail program.

Tasks:

1. Analyze the tape in terms of extemporaneous speaking criteria. That is, do you appear to be both prepared and informal?

2. Develop a plan to expand your five-minute explanation into a fifteen-minute presentation, complete with introduction, preview, visual aids, review, and question-and-answer period.

3. Ask your friend to tape your expanded presentation and then go over the two tapes with you. The two of you should focus on how well you maintained the extemporaneous style with the extra structure and responsibilities of the more formal presentation.

---

expected to prepare and distribute evaluation forms for your audience to fill out and return, either to you or to a sponsoring agent. You can then assess whether and how your presentation met your goals. In less structured situations, your analysis will have to be based upon much more impressionistic information: how many listeners asked questions, who requested further information, the ultimate action the group took as a result of your information, and the facial expressions of interest or disinterest. Whether the feedback is precise or ambiguous, however, remember that the main goal of evaluation is to do better next time, not to relive all the pain and agony (or joy) of a previous presentation. Feedback is not a time machine in which you travel backward through regrets, "should have dones," or celebrations; it is a forward-looking strategy for planning your next presentation and becoming a more successful professional.

GROUP PRESENTATIONS

Most of our reminders for individual presenters apply equally well to situations in which a team or group has prepared information for an audience. Obviously, some additional complications may arise in group presentations. For instance, team presentations may invite coordination problems and disorganization. Audiences can become confused by the intricate logistics of how group members juggle their materials and even wonder "Who's in charge here?" Teams preparing presentations must constantly ask themselves whether the advantages of the team's presence and interaction at the presentation outweigh the potential liabilities.

Team presentations are appropriate when a small group of two to four professionals has worked equally on a project, or when each person has brought a special expertise to the situation. They are especially appropriate, if not mandatory, for situations in which a question-and-answer session is likely to bring up different and specialized issues that no single presenter from the group could be expected to address with confidence. If a team presentation will best meet the needs of your task, you'll avoid some pitfalls if you remember to:

Divide the labor carefully. Ensure that your team's various tasks and how you divide them in the presentation make sense to listeners. For example, if there are four people in a presentation team, but one of them never speaks, the audience may wonder (rightly) why he or she is there. Audiences find such things distracting. If all four members are to participate, provide four meaningful roles for them to perform. As appropriate, briefly clarify the various presentational roles for the audience and the different areas of expertise represented by the team members.

Leave intragroup tensions at home. Participating in a business presentation means that all team members are cooperatively giving the *same* presentation. This is not like a classroom exercise in which participants receive different grades from a discerning professor, based upon their differing individual contributions. Few audience members will care who did what, as long as the outcome is good. Members of the team may have disagreed in the planning process about the content of recommendations or the ways they want to structure the presentation. This is natural. However, avoid advertising such dirty laundry to an audience that does not need to hear about it. Loud sighs, dismissive glances, shaking the head to signal obvious disagreement, and verbal put-downs of other team members will weaken the impact of any presentation. On certain extremely rare occasions, in groups with significant tension and content disagreement, a spokesperson might want to state calmly and simply that a particular issue "was difficult for us to decide, and not everyone in the group is fully convinced. However, on balance, we believe that . . . ." Only rarely will you need to identify who believes what, and even then you should stress the consensus decision making of the group.

Designate a primary spokesperson. Your primary speaker will typically take responsibility for introductions, transitions, summaries, and question-and-answer periods. A spokesperson need not be the formal leader of the group or even the person with the highest status within the organization, although these are the most likely possibilities because of audience expectations. The purpose of having a designated spokesperson, however, is pragmatic in that an audience needs to know whom to question for clarification and for insight.

We recommend that the spokesperson assume the role of coordinator rather than acting as the group's boss or authority figure. The differences may be subtle, but compare your own reactions to the two following statements: "I'd now like to call on Jim, who will provide you with what I think is our strongest recommendation"; and "Our next speaker is Ramona, who did most of the work on the especially important Jackson Glove account." The first introduction spotlights the introducer as the leader, while the second clearly indicates the extent of Ramona's contribution.

Interrupt only to crisis-manage. Few aspects of team presentations are more irritating to audiences than when presenters tend to talk over each other's words. Some overlapping of speech is normal in any conversation, of course, and

EFFECTIVE PRESENTATIONS

INDIVIDUAL

- Plan for spontaneity.
- Sensitize yourself to nonverbal cues.
- Introduce with clarification of your credibility.
- Preview the presentation.
- Stay with organization.
- Review for impact.
- Stress dialogue.
- Follow up with analysis of feedback.

GROUP

- Divide labor.
- Leave intragroup tensions at home.
- Designate a primary spokesperson.
- Interrupt only to crisis-manage.
- Practice transitions and team roles.
- Do a full rehearsal.

parts of most team presentations will become appropriately conversational. However, constant interruptions will give the impression that your group doesn't know what it is doing. That said, your group should still remember that some interruptions are necessary—for instance, to correct a misstatement of an important statistic. Ideally, the spokesperson would be the one to interrupt in such cases.

Practice transitions and team roles beforehand. In many situations, we have observed groups of students and executives working well together in an extended project. Surprisingly, many of these groups assume that their work is finished when they have decided on content and have divided up the roles for their presentation. Members then each go their separate ways until the appointed time when they reconvene, perhaps five minutes before the board meeting, to find that they have not decided even how to pass the conversational baton to each other within the presentation. The awkward and embarrassing moments this creates could be avoided by an hour or so of rehearsal the day before. At the rehearsal the group runs through its plan and makes final decisions about such matters as who goes first, how to set the stage for the next speaker, how equipment needs are being handled, and how each person's responsibilities for content might affect what others will say.

If possible, do a full rehearsal. A full run-through allows a team to make needed adjustments and make contingency plans in case the unexpected happens. For example, one part of the presentation might be too long or too short. Perhaps one presenter is using technical jargon the audience might not understand. While diverse styles are often effective, the whole presentation must hang together. You can also decide at this stage whether your audiovisual aids are appropriate and who will run equipment, distribute handouts, or change overhead transparencies during the presentation. Therefore, at least one full dress rehearsal is useful.

Summary

Accountants, like Francine Marti and her team, often find themselves in a variety of oral communication contexts and need to anticipate how these situations can affect their credibility within an organization. Perhaps you hope you will be evaluated exclusively on the basis of your professional accounting expertise, your facility with computers, your analysis of complex trends, or your ability to project trends into the next decade. Rightly or wrongly, however, others tend to form and maintain their impressions of you based largely on how you conduct yourself in presentations and interpersonal relationships.

This chapter has not surveyed every possible oral communication context in which you will find yourself. Instead, we have concentrated on helping you overcome the problems raised by communication apprehension, interview and listen to others meaningfully, and improve your presentational speaking style individually or in groups.

Practicing and mastering these skills can increase your self-confidence and boost your career potential. Remember that when your company decides who to promote to that recently opened supervisory position, it will not be just your accounting skills they'll evaluate. They will be looking for a credible communicator, someone who interacts effectively in face-to-face situations.

How to Approach Practice Tasks

ANALYZING NORMAL AND ABNORMAL NERVOUSNESS

Many people won't be able to recall readily many details about their past three speeches or presentations. Sometimes we're so relieved when the presentation is over that we block out the memories for a time. However, with a little effort, you should be able to construct fairly lengthy lists of such personal behaviors as the following: arrived early to speak, met many new people ahead of time, arranged chairs and tables in a semicircle, made sure my watch was visible in front of me as I spoke, switched on more lights at the front of the room, and so on. Symptoms of nervousness you might recall could include, among other things, noticeable heartbeat, sweating, dry mouth, inability to concentrate on small talk before the presentation, rapid speech, inability to pronounce some words fluently, and lack of eye contact with the audience.

In analyzing links between nervousness and speaking situations you might conclude, as many do, that you were less nervous when you took control of many aspects of context proactively, instead of leaving them to chance or the whims of others. You are not a passive victim of presentational expectations, but you can influence the action with positive consequences. If you do so, you'll probably be less nervous.

PRACTICING AN INTERVIEW

In these tasks, you are imagining yourself in the role of interviewer and looking for possible misunderstandings that could have been cleared up by perception checks and active listening. In a 5- to 10-minute interview, there may be as many as five or six occasions where a skillful perception check would probably produce clearer understanding of the expert's message. Your perception checks should resemble these in form:

> "Did I hear your figures accurately? I think you meant to predict a 4.7 percent increase, but it almost sounded like you said '47.' Did I get it wrong?"
> "Some viewers may not know how the Federal Reserve Board is comprised, even though they've heard of it. Did you suggest that . . . ?"

ANALYZING AN ORAL PRESENTATION

You should view your tape critically for evidence that you can sound confident and prepared even while speaking conversationally with an audience. Conversational speaking usually is indicated by such things as ready eye contact (without staring), significant vocal variety, comfortably animated gestures, easy transitions between points, and a somewhat informal language style that avoids slang which could demean your professionalism.

When your explanation is expanded into a 15-minute presentation, make sure it is clearly organized. Some beginning speakers think that *extemporaneous* means "off the cuff" or "winging it," without a specific predetermined organization. Rather, extemporaneous speaking refers to a style that is conversational and spontaneous-sounding *and* carefully planned. Your presentation should exhibit a clear and engaging introduction that grabs audience attention, a preview of the main idea and specific points, development of three or four basic ideas, a summary, a motivational closing, and a clear transition to a question-and-answer period. You and your friend/critic should look especially for how well you retained your informality and friendly accessibility while meeting the demands of a more complicated presentational structure.

Endnotes

1. R. Hoff, *"I Can See You Naked": A Fearless Guide to Making Great Presentations* (Kansas City, MO: Andrews and McMeel, 1988), pp. 30, 138. Our description here of the difference between presentations and speeches is based on Hoff's insightful discussion.

2. L. Zunin, *Contact: The First Four Minutes* (New York, NY: Ballantine, 1972).

3. For example, see R. Nelson, *Louder and Funnier: A Practical Guide for Overcoming Stagefright in Speechmaking* (Berkeley, CA: Ten Speed Press, 1985).

4. See G. M. Killenberg and R. Anderson, *Before the Story: Interviewing and Communication Skills for Journalists* (New York, NY: St. Martin's Press, 1989); and R. Gorden, *Basic Interviewing Skills* (Itaska, IL: Peacock, 1992).

5. See M. Purdy and D. Borisoff (Eds.), *Listening in Everyday Life: A Personal and Professional Approach* (2d ed.) (Lanham, MD: University Press of America, 1997); and A. Wolvin and C. G. Coakley, *Listening* (5th ed.) (Dubuque, IA: Brown & Benchmark, 1996).

6. M. Holcombe and J. Stein, *Presentations for Decision Makers: Strategies for Structuring and Delivering Your Ideas* (3d ed.) (New York, NY: Van Nostrand Reinhold, 1996).

7. R. Hoff, pp. 3–7.

8. T. Leech, *How to Prepare, Stage, and Deliver Winning Presentations* (New York, NY: American Management Association, 1982), pp. 4–6.

Index